W9-BAV-442

From Quarks to the Cosmos

FROM QUARKS TO THE COSMOS

Tools of Discovery

Leon M. Lederman

David N. Schramm

SCIENTIFIC AMERICAN LIBRARY

A division of HPHLP
New York

Library of Congress Cataloging-in-Publication Data

Lederman, Leon M.
    From quarks to the cosmos.

    Includes index.
    1. Astronomy.   2. Particles (Nuclear physics)
3. Science—Philosophy.   I. Schramm, David N.   II. Title.
QB43.2.L43 1989   520   89-10207
ISBN 0-7167-5052-X

Copyright © 1989 by Scientific American Library

No part of this book may be reproduced by any mechanical,
photographic, or electronic process, or in the form of a
phonographic recording, nor may it be stored in a retrieval
system, transmitted, or otherwise copied for public or private
use, without written permission from the publisher.

Printed in the United States of America.

Scientific American Library
A Division of HPHLP
New York

Distributed by W. H. Freeman and Company
41 Madison Avenue, New York, New York, 10010 and
20 Beaumont Street, Oxford OX1 2NQ, England

1 2 3 4 5 6 7 8 9 0   KP   7 6 5 4 3 2 1 0 8 9

This book is number 28 of a series.

To Ellen and Judy

CONTENTS

PREFACE

All thinking people in their quiet moments, ask themselves the eternal questions: Why am I here? What is the reason for all existence? When gazing in awe at an unusually clear night sky, they ask: What is up there? How did the universe happen? And why is it so beautiful? These questions have always been at the center of religious, philosophical, and scientific thought. Our intention in this book is to reveal the strategy by means of which modern physics is constructing a great intellectual edifice—nothing less than a complete understanding of the world—from the vast scope of the creation of the universe to the microworld inside the atom where the laws of nature are understood.

This is a story about the unimaginably small and the incredibly large. It is a story about the inner space of atoms and what they are made of, and it is a story about outer space, about planets and stars and galaxies, in fact about the entire universe.

The two stories of inner space and outer space come together in the 1980s in a most dramatic way. Indeed, that merger has been reflected in our professional lives. One of us (David Schramm) is a theoretical astrophysicist; the other (Leon Lederman) is a particle experimenter turned laboratory director. The laboratory is the Fermi National Accelerator Laboratory, where, in 1989, the highest-energy particle collisions in the world take place. Back in 1981, while hiking together in the Dolomite mountains of Italy, we realized that the deep connection between particle physics and early universe cosmology could be more fruitfully explored by embedding a theoretical astrophysics group within the accelerator laboratory. After obtaining initial funding from the National Aeronautics and Space Administration, we worked together in collecting a young and dynamic group of astrophysicists, whose seminars and lectures helped to stoke the intellectual fires of the laboratory. The "astros" were in turn stimulated by the detailed information on subnuclear phenomena, which they badly needed to model the evolution of the universe from its early stages, when it existed as a soup of quarks and leptons. The success of this unique collaboration encouraged us to write the book you have in your hands. We would especially like to acknowledge the Fermi Laboratory and its patron, the U.S. Department of Energy, for providing the locale, the atmosphere, and the excitement of intellectual engagement, so crucial to the writing of this story.

Although many of the ideas that emerge from the inner space/outer space merger are abstract, they are firmly based in experimental science. We hope to convey the intimate relationship between experiment and theory that has led to the new understanding of the universe. That relationship is the very heart of what is called the scientific method. Clearly it is a deep subject, but we try to avoid philosophical subtleties and keep to the essential points in order that the reader can better enjoy the inner space/outer space adventure.

Because they are the core of experimental science, we decided to focus especially on the experimental tools, whose evolution over the past fifty or so years is by itself as remarkable as the depth of the theoretical synthesis. The "telescopes and microscopes" story tells the development of particle and radiation detectors—from the earliest Geiger counters to the latest silicon strip microvertex devices, from the bunsen burner to the superconducting supercollider, and from Galileo's hand-held telescope to the huge radio telescopes and, soon, the flight of the Hubble Space Telescope.

Finally, we appeal to the poet Victor Hugo, who asked, "Where the telescope ends, the microscope begins. Which of the two has the grander view?" Only by looking at both the very small and the very large can one hope to understand the universe.

Leon M. Lederman
David N. Schramm

April 1989

From Quarks to the Cosmos

1 AN INTRODUCTION TO INNER SPACE AND OUTER SPACE

We begin by telling two stories: one about the world of the very small, the other about the universe around us.

A STORY OF INNER SPACE

On April 17, 1987, the control room at Fermilab's Tevatron accelerator was unusually crowded for 2 A.M. The seven-man operating crew, helped by a half-dozen experts, were watching their computer screens—checking the color graphics, rolling the cursor controls, and reading data on the operation of the world's highest-energy accelerator. They reviewed a seventeen-item checklist in a process like a somewhat amateurish NASA countdown. A very dense stack of antiprotons (called p-bars), the negatively charged antimatter twin of protons, had been accumulated during the previous twenty hours and were speeding around the 1,700-foot-circumference accumulator storage ring, waiting for transfer to the Tevatron. If the transfer of antiprotons was successful, two beams of particles, one of protons and one of antiprotons, would circulate in opposite directions in the Tevatron's racetrack-shaped vacuum pipe, resulting in head-on collisions.

The power of an accelerator to explore new domains smaller than the atomic nucleus depends on two factors: (1) the energy to which it can accelerate particles, and (2) the number of particle collisions per second

On February 23, 1987, light from this exploding star first reached Earth from its origin in the Large Magellanic Cloud 170,000 light-years away. The supernova ejected particles called neutrinos, whose detection on Earth was made possible by a new and close interplay between astronomy and particle physics.

The main control room at the Fermilab accelerator on a quiet day.

that can take place within it—a number that physicists refer to as its *luminosity*. The Fermilab collider had reached the highest energy of any accelerator in the world, but its luminosity had been too low. The test on this crucial night was going to be an all-out attempt to breach the intensity barrier that had slowed research and frustrated the experimenters for the past several months. The collider had to prove its ability to deliver a significant number of collisions per second.

"P-bar shot in thirty seconds."

The manipulation of the antiprotons, whirring around the accumulator ring at a velocity close to the speed of light, required meticulous precision in the computer-based controls of hundreds of instruments: bending and focusing magnets, deflection "kicker" magnets, radiofrequency accelerators, and sensor devices too numerous to mention. After all, antimatter, discovered in the 1950s and the grist for so much science fiction, was still a fairly scarce and exotic species of nuclear particle. There was still a big chance the experimenters might fail and lose the precious store of antiprotons.

The crew sat poised at their consoles as the button was pushed. The antiprotons shot from the storage ring to the Tevatron. In the accelerator control room, there was a gasp of relief. But in the experimenters' control room, half a mile away, a bedlam of cheers rose up as the physicists observed the densest bunch of p-bars ever to enter the Tevatron, achieve 900 billion electron volts (GeV) of energy, and begin to collide head-on with protons moving in the opposite direction.

The path of the protons (red) and antiprotons (green) in the two rings in the tunnel of the Fermilab Tevatron. The diameter of the rings is 2 km (1.3 mi), and 20 inches separate the two rings vertically. The Linac accelerates protons to 200 MeV; they are then accelerated to 8 GeV by the booster before being injected into the main ring.

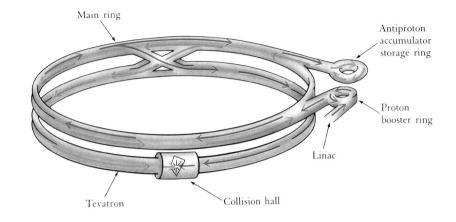

The collision rate jumped visibly. "One point three, ten to the twenty-ninth," called one physicist, and another cheer broke out. Physicists started arriving from other areas of the accelerator complex to watch the computer display of 1,800 GeV collisions, coming in at a rate of 5,000 a second. Someone produced a case of warm domestic champagne and styrofoam cups appeared. The overjoyed physicists began this spontaneous middle-of-the-night party to celebrate the knowledge that ten years of incredible work on the part of hundreds of scientists, engineers, and technicians had not been wasted. The experiment worked. A new domain within the atomic nucleus was open for exploration.

However, the real action was taking place about 100 yards below the experimenter's control room in the collision hall. At opposite ends of the hall two stainless steel tubes, about 4 inches in diameter, emerge out of dark tunnels. These tubes join in the middle of a three-story-high detector complex. Bunches of protons the size of soda straws flash through one highly evacuated tube, going east to west; antiproton bunches move through the other tube with equal speed from west to east. As they approach the midpoint, these bunches are squeezed narrower and narrower by powerful focusing magnets, from the thickness of a soda straw down to that of sewing thread. The bunches pass through one another, seemingly unscathed—but not quite. The particles are crowded together to increase the probability of a collision. About every tenth pass, a single proton in one dense swarm collides head-on with a single antiproton in the other.

The collision is impressive. New particles are created by the energy released in the collision and scatter from the interaction point in all directions. Minute electronic wakes generated by the emerging particles are amplified and recorded by thousands of sensors in the massive detector that surrounds the interaction point. Each particle is registered in exquisite detail.

In the experimental control room, a graphic display screen followed instructions from the detector's computer, and began to sketch the event in a manner simplified enough to be comprehensible to the assembled Ph.D.s. After a cursory examination of the nuclear debris, the very selective computer rejects most of the collisions as being of "routine interest only." But several times a second an event is deemed interesting enough that some 80,000 channels of information are recorded on the magnetic tape, the indelible memory of the experiment, for further analysis. Within these selected events the physicists suspect they have captured a head-on collision of the quark constituents of the colliding protons and antiprotons.

The colliding detector at Fermilab, or CDF, is the crown jewel of U.S. high-energy physics. It weighs 5,000 tons, but is instrumented with the delicacy and precision of a Swiss watch. Some months earlier, this three-story-high device was moved at the rate of one foot per hour from its assembly garage into the interaction hall, trailing an umbilicus of thousands of cables that would carry data from the detector to its computer controllers. Some 80,000 electronic channels of information are collated by dedicated computers, examined, filtered, and ultimately read onto magnetic tape for future analysis. The detector represents the labor of about 300 scientists, students, engineers, and technicians, working for almost eight years.

The three-story-high colliding detector at Fermilab. The black arches, which have been removed to the side, contain instruments for measuring particle energies from collisions.

The accelerator and the antiproton source device are similarly the results of ten years of designing, inventing, and building to create a collection of instruments that are well beyond the state of accelerator art as envisioned in 1978. Taken together, the detector, accelerator, and antiproton source device constitute a tool of incredible power for advancing the knowledge of the microworld, the subnuclear domain of the infinitesimal. The proton–antiproton collider at Fermilab was built to provide experimental data needed to arrive at an all-encompassing and unified theory of the structure of matter and energy, space and time.

At the time the Tevatron was built, the state of particle physics was described by an elegant synthesis known as the standard model. This held that all matter is composed of two kinds of particles, *quarks* and *leptons,* which interact and cluster together via the agency of four forces: *strong, weak, electromagnetic,* and *gravitational.* The Tevatron was built with two general goals in mind: to better measure the properties of the particles and forces as described by the standard model; and, via the collider, to go beyond the standard model by providing data on the correct road to unification of the forces and simplification of the standard model particle structure.

The density of energy concentrated in the collision volume is stupendous. Upward of fifty newly created particles race out from a cauldron in which almost 2,000 billion electron volts (GeV) of energy boil up

A computer reconstruction of a particle collision at Fermilab. This central tracking chamber plot shows a beam's eye view of the many particles emerging from a proton–proton collision at 1.8 trillion electron volts.

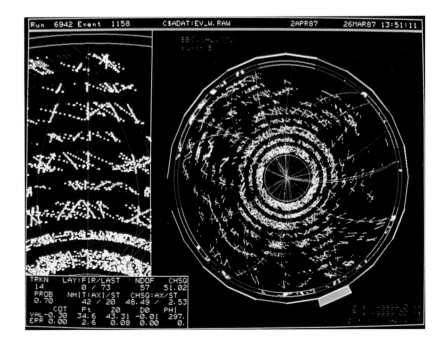

from a region of space so small that 1,000 billion of these regions would fit into the nucleus of the hydrogen atom. In that tiny spot, for the very briefest instant, thirty feet below a former corn field in Illinois, the experimenters have replicated the conditions that existed throughout the universe some billionths of a second after its creation in the cosmic explosion known as the "big bang," 15 billion years ago. It is here that we begin to perceive the relationship between particle physics and cosmology that will be the theme of this book.

A STORY OF OUTER SPACE

Deep in two underground mines on opposite sides of the globe, unprecedented events were taking place on February 23, 1987, at precisely 7:35 A.M. Greenwich mean time (GMT).

Several years earlier, enormous plastic tanks (bigger than Olympic-sized swimming pools) had been installed in the Morton Salt Mine under Lake Erie near Cleveland and on the other side of the world in the Kamioka Zinc Mine north of Tokyo. The tanks had been filled with ultraclean water and outfitted with thousands of photomultiplier tubes,

A physicist is making repairs inside the underground particle detector located in the Morton Salt Mine near Cleveland, Ohio. The box measures about 100 feet on a side and is filled with very clean water. The walls of the box are lined with photomultiplier tubes, sensitive to extremely minute flashes of light.

Top: Part of the Large Magellanic Cloud as it appeared to us before Supernova 1987A. *Bottom:* The same area of space after Supernova 1987A. The exploding star is clearly visible in the lower left.

which detect minute flashes of light in the dark mine water. Whenever a tube detects a flash, it sends an electronic signal to a computer elsewhere in the mine. This computer records the timing and location of the flash. Down inside the mine, the tank is shielded not only from the Sun's light (and that of the Moon and stars), but also from the cosmic rays that would trigger occasional flashes if the detector were on the Earth's surface.

These detectors had been operating for several years, searching for the flashes that would indicate the possible spontaneous decay of protons in the water. They had seen none. However, on February 23, at precisely the same moment, the Kamioka detector recorded eleven separate events and the Morton Salt Mine detectors recorded eight. Each event consisted of about thirty photomultiplier flashes occurring in a particular time sequence and pattern, implying that something had entered the detector and interacted. All of these events occurred in about ten seconds. Never before had either of these detectors recorded more than one or two random background events within ten seconds. Why should these two detectors, far from each other and deep underground, suddenly experience similar bursts of activity of an unprecedented kind?

That night, on a mountain top in Chile, a Canadian astronomer, Ian Shelton, was using a small telescope to photograph the sky. For several nights Shelton had been taking pictures of a clump of stars known as the Large Magellanic Cloud, a small galaxy orbiting our own Milky Way at a distance of about 170,000 light-years away. (A light-year is the distance that light travels in one year—9.46×10^{15} m.) To Shelton's surprise, his new photographic plate had a big bright dot on it that wasn't in any of his previous photos. At first, he thought a piece of dust or something else had messed up his plate. Considering the size of the dot, Shelton knew that if it showed a real astronomical object, he should be able to see it even without his telescope. So he went outside and looked. There it was, Supernova 1987A, the first exploding star visible to the naked eye in 384 years.

After Shelton notified the world's astronomical community, people looked at other photographic plates of the Large Magellanic Cloud taken around the same time. The European Southern Observatory found that at zero hour (midnight GMT) on February 23, they had made a photographic plate showing all the stars still there, with no supernova. At 9 A.M. (GMT) on the 23d, an amateur observer in New Zealand named Albert Jones was looking at the Large Magellanic Cloud, as he did almost every night. He hadn't noticed anything special either. But by 11 A.M. (GMT), a satellite tracking camera in Australia had a photo showing a new object a million times brighter than our Sun. (The next night, without knowing of Shelton's discovery, Jones independently found the supernova.) Hence light from the exploding star first reached Earth after 9 A.M. and before 11 A.M. on February 23.

More than twenty years earlier, theoretical astrophysicists had argued that a supernova explosion should release a huge number of subatomic particles called neutrinos. Neutrinos had been studied at particle accelerators and reactors, and they were known to travel at or near the speed of light and to interact so weakly with any other particles that they easily pass through the Earth. The neutrinos would be emitted from the very center of the exploding star, and would escape right through the overlying material. The shock wave from the central explosion would, after a few hours or days, make its way to the surface of the star and produce a blast of light as the star flew apart.

All the pieces fit together. At 7:35 A.M. Greenwich mean time on February 23, neutrinos from the explosion itself reached Earth, triggering the huge underground detectors. Several hours after the initial explosion, the shock wave had reached the surface of the star and produced a blast of light, which then was seen on Earth several hours after the neutrinos. All of the action actually took place 170,000 years ago in the Large Magellanic Cloud, but the wave of neutrinos and light took that long to reach us. In order for the observed number of events in the underground detectors to have occurred, more than 10 billion neutrinos must have hit every square centimeter on Earth. Obviously, most passed right through without notice, but the underground water detectors were sensitive enough to pick up nineteen of them.

The deciphering of the supernova demonstrates the new and close interplay between astronomy and elementary particle physics—the scientific field where things such as neutrinos are studied. The kind of detectors used in these mines are of the style developed for studying reactions at high-energy accelerators, and they were built to test particle theorists' ideas about the nature of matter. But the experiment's greatest success was the birth of a whole new branch of science known as "neutrino astronomy."

CONNECTIONS

These two stories about inner and outer space reveal the merging of two scientific disciplines that could hardly appear to be more distinct. Particle physics seeks to find a simple and orderly pattern to the behavior of matter on the atomic and subatomic level. To this end large particle accelerators are built, acting, in effect, like giant microscopes that zoom down through the atom, down into the nuclear core, and into the internal structure of the proton and neutron denizens of the nucleus. Here we seek pattern and order—the basic objects and laws of nature applicable

The horizontal, almost straight tracks were made by a beam of 300-billion-electron-volt protons from the Fermilab accelerator as they passed through a 30-inch liquid-hydrogen-filled bubble chamber, a type of particle detector. One proton collides with the nucleus of hydrogen and, in the collision, 26 charged tracks are produced. The entire chamber is in a powerful magnetic field that curves the tracks.

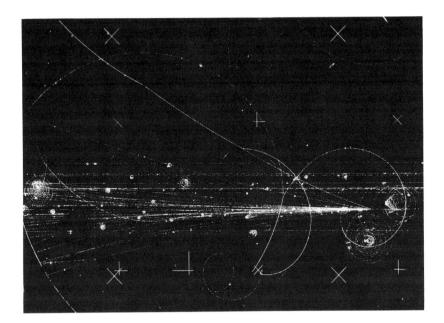

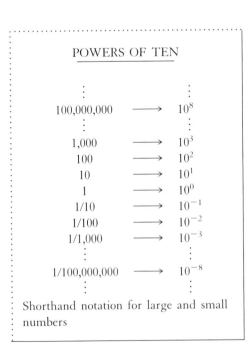

POWERS OF TEN

100,000,000	$\longrightarrow$	10^8
1,000	$\longrightarrow$	10^3
100	$\longrightarrow$	10^2
10	$\longrightarrow$	10^1
1	$\longrightarrow$	10^0
1/10	$\longrightarrow$	10^{-1}
1/100	$\longrightarrow$	10^{-2}
1/1,000	$\longrightarrow$	10^{-3}
1/100,000,000	$\longrightarrow$	10^{-8}

Shorthand notation for large and small numbers

in this realm of the unimaginably small. The distance scale, here magnified by the complex accelerators, is impressive. Human dimension is one meter (10^1), the length of an arm; the nucleus of the hydrogen atom is 10^{-15} meters; and the Tevatron focuses down to one hundredth of this scale (10^{-17} m)—that is, inner space!

Astronomers build equally complex devices—telescopes and observatories. These gather data from our solar system (10^{11} m wide), from our spiraling Milky Way Galaxy (about 10^{20} m across), from distant galaxies and clusters of galaxies, all the way out to the rim of the cosmos, beyond which light has not yet reached us (some 10^{26} m or 15 billion light-years away)—that is, outer space! Astrophysicists interpret the data from these telescopes in order to properly describe the universe around us. But, like archaeologists, they also read, from the drama of position and motion, the story of the origin, pattern of evolution, and natural laws that govern this vast domain.

We are seeing here a convergence between particle physics and that subdiscipline of astronomy dealing with origin and evolution—early universe cosmology. The instruments, and even the stated objectives, are different, but the languages draw ever closer. And, as we shall see, the laws of nature that control and order the microscopic world, and those that determined the creation and evolution of the universe, from its incandescent birth to the luminescence of a winter night's sky, are beginning to look identical.

This merging is relatively recent and only recently intense. However, at the dawn of science, about 2,500 years ago in the Greek cities along the Aegean, inner and outer space were both essential and interconnected domains. The Milesian philosophers believed in overarching principles of poetic simplicity that gave coherence and order (*kosmos*) to an apparently complex world. The atoms and the voids, the consideration of primary substances and the structure of the heavenly vault, were a single subject in those times. During the past few hundred years, the subjects separated, constructed their very different instruments, and made their separate discoveries. Some overlaps persisted. Progress in nuclear physics in the 1930s explained why the Sun continues to burn, and cosmic radiation from astronomical sources was used to search for new particles and behaviors. But the real convergence began in the 1970s. In the closing years of the twentieth century, we find again a unity of knowledge like that envisioned by our ancestors. It is this new unity achieved by mind and instrument that we will address in this book. As we examine how the study of the very large (the cosmos) is merging with the study of the very small (fundamental particles), we will especially focus on the tools that are used to study both the macro and the micro worlds.

In the next chapter, we will begin exploring the origins of modern physics and the development of our current views on the nature of space, time, matter, and forces, and we will show how the invention of new instruments influenced the development of new ideas. We will discuss the early unification of electricity and magnetism that occurred through the work of Faraday, Maxwell, and Hertz. By unification we mean the discovery of an underlying principle that enables two or more apparently separate forces to be described as the action of a single entity. We will also discuss how our entire view of nature was changed by Einstein's and Bohr's revolutionary ideas of relativity and quantum mechanics, and then go on to focus on a major crisis in present-day understanding—how to unite quantum mechanics with general relativity (gravity theory).

In Chapter 3, we will see how scientists began to probe into the heart of the atom through the use of particle detectors, how the exploration of the nucleus of the atom first identified the neutron and the proton, and how one nucleus can be transmuted into another. This transmutation gives us some understanding of how heavy elements could be built up from lighter ones (natural "alchemy"), and of how energy might be generated in stars by these nuclear reactions. Thus, the study of atomic nuclei solved a long-standing astronomical question: What makes the stars shine?

In Chapter 4, we will see that neutrons and protons are not really the elementary building blocks they were once thought to be, but are made out of smaller particles called quarks. These particles and their interactions are explored using very large accelerators that now measure many miles around. By the end of Chapter 4 we will have developed what physicists call the standard model. This model is a summary of our current understanding of the structure of matter, its particles and forces, and their interrelationship.

In Chapter 5, we will see that, while this understanding of matter and forces was developing, important new things were being learned on the cosmic scale as well. Astronomers were no longer restricted to viewing stars and galaxies in the narrow range of optical light, but with the help of satellites, balloons, enormous dish antennas, and other devices, they were able to "see" these objects by detecting the energy they emit, ranging from gamma rays to radio waves, as well as neutrinos. With these observations astrophysicists were able to piece together a much better understanding of what these different objects in the universe are and how they might have come into being.

Chapter 5 will also discuss the emergence of the universe from the intense heat and denseness that we call the big bang. The early universe was very hot indeed, achieving energies that can now be reached only in elementary particle accelerators. Thus, we can use particle accelerators to understand the earliest moments of the universe. The ability of accelerators to mimic the early universe has led to the unification of the study of the very large and the very small—the merger of inner and outer space.

In Chapter 6, we will explore the connection between inner and outer space, showing how physicists hope to understand the whole picture, from clusters of galaxies down to the smallest particles of matter, through a single theory—sometimes called a "theory of everything" (TOE). Current ideas about TOEs might help us to understand the origin of objects in the universe, as well as the structure of matter itself, perhaps even answering such abstract questions as how space and time came about, and how our universe came to have three dimensions.

In Chapter 7, we will look at the tools that will be available in the 1990s for probing these TOEs, to determine whether these theories are the right ones, or if some new, unthought-of concept will be needed to explain everything. These tools will include such things as the Superconducting Supercollider—an accelerator, 52 miles in circumference, that is planned for a site in Texas. Other tools for the 1990s include a space telescope that can "see" in optical, infrared, and ultraviolet light, and new space observatories to detect and analyze X-ray and gamma-ray radiation. In addition, there will be giant neutrino observatories that, instead of going into space, will be built deep underground, to shield out everything but the neutrinos and will provide clues about exotic objects in the universe.

In the epilogue, Chapter 8, we will attempt to provide a philosophical perspective for these developments.

However, before we proceed with our story, we'd first like to give you a brief glimpse of the modern view of both matter and the cosmos.

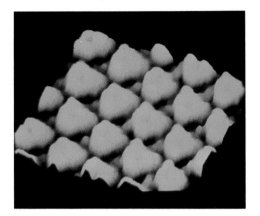

Six carbon atoms in a ring form a benzene molecule, revealed here by modern techniques of electron microscopy.

THE MODERN VIEW OF THE NATURE OF MATTER

The modern view of atoms and molecules and the particles they're made out of only started to come into being at the beginning of this century. Unlike the ancient Greeks, who observed what was taking place in their world, thought carefully about it, but had no way to test their conclusions, the modern view of the structure of matter is based on many decades of experiments checked and duplicated the world over. These experiments showed the truth of the Greek idea that all matter could be broken up into small particles which we, following the Greeks, call "atoms." Also, we learned that there are about 100 kinds of these atoms occurring in nature.

Atoms combine to make simple molecules, such as water, or complex organic molecules such as DNA (dioxyribonucleic acid), the prime building block of all life on Earth. DNA is made up of millions of atoms.

The atom is largely empty space, although occupied by a cloudlike swarm of orbiting electrons, each one carrying a tiny negative electrical charge. The core of the atom is known as the nucleus, and it is about 100,000 times smaller than the entire atom. To understand this scale, think of a very large auditorium, perhaps your favorite baseball stadium. If an atom were magnified to the size of this stadium, the nucleus would be represented by a grain of rice, floating in the center.

Atoms bind together to make molecules by sharing electrons. The nucleus, where the bulk of the mass is located, does not really enter into the chemical reactions that make up molecules. The nucleus is made out of neutrons and protons. The protons carry a positive charge. The neutron has no electric charge whatsoever, but its mass is comparable to that of a proton, which is approximately 2,000 times the mass of an electron. The number of protons in the nucleus determines the number of electrons in the outer part of the atom, because each positively charged proton must be offset by a negatively charged electron for the atom to be electrically neutral. Thus, we can understand that through these electrical forces electrons are held in their orbits around the nucleus of the atom. Since atoms themselves are not too big (radius about 10^{-8} cm = 10^{-10} m), and only the largest can be seen even with the most powerful electron microscopes, it is obvious that the nucleus is a very tiny thing indeed, which remains invisible. In fact, to "see" the nucleus requires accelerators, which extend the limits of conventional microscopic sight.

Until recently, it appeared that all 100 or so atoms could be understood as combinations of three particles: the neutron and the proton making up the nucleus, and the electron making up the outside. Thus,

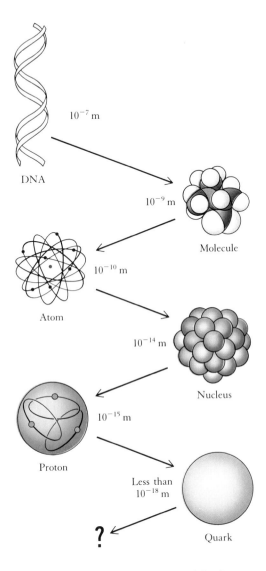

10^{-7} m

DNA

10^{-9} m

Molecule

10^{-10} m

Atom

10^{-14} m

Nucleus

10^{-15} m

Proton

Less than
10^{-18} m

Quark

?

The successive "seeds within seeds" of matter, from DNA to quarks. Note the change of distance scale at each level.

this messy bunch of 100 different kinds of material could all be reduced to three basic building blocks of nature. However, we discovered that this idea was not the final word. In the 1950s and 1960s, experiments were done at higher and higher energies, taking advantage of the existence of new and very powerful particle accelerators (occasionally called "atom smashers" in the popular press). In the subsequent probing of the neutron and proton, a whole zoo of new particles were found. Following the ideas that led to the reduction of 100 atoms to only three fundamental particles, physicists suspected that this new, huge number of particles really indicated that even smaller, more fundamental particles existed. Experiments in the 1970s proved that three smaller particles called *quarks* could be combined to make up neutrons, protons, and many of the multitude of other particles. Although the neutron and proton are no longer fundamental, the electron (now considered a member of a larger family called *leptons*) appears to be as basic as the quarks.

Any discussion of these fundamental particles leads inevitably to an examination of the forces that bind them together. We've already mentioned the *electromagnetic force* that holds the electrons and nucleus together in the atom. But there are three additional forces. As we indicated earlier, these are part of the theoretical system known as the standard model.

In addition to interacting via the electromagnetic force, quarks seem to interact via another, stronger force, which we call the *strong interaction*. This strong binding of the quarks explains how neutrons and protons can be packed closely together without the nucleus blowing apart from the opposing electrical charges of the protons in it. The strong force fields inside protons and neutrons reach out sufficiently to bind the protons and neutrons firmly into the nucleus.

We are all familiar with forces that push or pull, that attract or repel, but there are also forces that change one particle into another. Collectively we call all of these *interactions*. There appears to be an interaction which changes one kind of quark into another kind. This is called the *weak interaction*. Thus, matter seems to require three kinds of interactions to behave as it does: electromagnetic, which holds the electrons to nuclei; the strong force, which holds the quarks together; and the weak interaction, which can change one kind of quark into another, or, equivalently, can change a neutron into a proton or a proton into a neutron.

The standard model includes one more force. Although it doesn't play a very important role in the make-up of atoms and molecules, it is essential to the structure of the universe itself. It is gravity—the force that draws any two objects with mass toward each other. Because the masses of single atoms are so small, the gravitational force is negligible at the atomic level. At this level the three forces mentioned above are much more important. (Even the weak force is much, much stronger than gravity.) However, when we start talking on the cosmic scale, we find that gravity is the dominant force in the universe—it moves the stars and planets and holds the very universe together.

The reason the gravitational force wins out on the large scales, while losing out on the small, is that each of the other forces ends up canceling out, so that they are not applicable at large distances. For example, the electromagnetic force requires electrical charges for there to be either attraction or repulsion. However, atoms tend to be electrically neutral; so large clumps of atoms will have no net electric charge. Similarly, nuclei seem to be quite satisfied as nuclei, and don't interact strongly outside the domain of the nucleus. Likewise, the weak force, which changes quarks, only applies within the domain of the nucleus. Large clumps of atoms do not show any exterior strong, weak, or electromagnetic force; the only force left, that is cumulative, is the extraordinarily feeble force known as gravity. When enough atoms are clumped together, as within planets and stars, that force becomes dominant.

THE MODERN VIEW OF PLANETS, STARS, AND GALAXIES

As with the structure of matter, the modern view of planets, stars, and galaxies is based on repeated observations made by people the world over and throughout history. However, unlike studying atoms and molecules, which is done in the laboratory where we can control the environment, we can study astronomical objects only by looking at the radiations that emanate from them and arrive at Earth or at sensors, such as satellites and space probes. We can't manipulate a star to see how it works; we must, instead, try to find stars in varieties of conditions and states of development. In this sense, astronomy is more like the ancient Greek approach to science—observe and try to understand. One striking difference is that modern astronomers have vastly more powerful means of observing. The other is that the laws of physics, which have been developed largely on the basis of terrestrial experiments, provide astronomers with a basic framework for understanding astronomical phenomena.

It has been known since the sixteenth century that our local region of the universe has a large central object, known as the Sun. Our Sun is orbited by certain planets, Mercury, Venus, (Earth), Mars, Jupiter, Saturn, that are visible to the naked eye. The planets, unlike the Sun, do not radiate visual light, but are seen only by the reflected light of the Sun. Telescopes enable us to find fainter objects in the sky, and they have been

The plane of the Milky Way stretches across the night sky.

used to discover more distant planets: Uranus, beyond Saturn; Neptune, beyond Uranus; and a ninth object, Pluto. This basic picture was discussed by Copernicus and worked out more thoroughly by Johannes Kepler, who utilized the detailed observations of planetary positions made by Tycho Brahe. The planetary motions were subsequently accounted for in mathematical detail by Newton's laws of motion and his theory of gravitation. The realization in the sixteenth century that the Earth is not at the center of the solar system, and not at the center of the universe, was one of the philosophic ideas that contributed much to the modern rethinking of the world that began during the Renaissance.

We also know that our Sun is a star like the approximately 100 billion other stars in our Galaxy, the Milky Way. The Sun has a mass of 2×10^{33} grams $= 2 \times 10^{30}$ kilograms $= 2 \times 10^{27}$ tons. By contrast, the Earth has only one millionth the mass of the Sun, and the largest planet, Jupiter, only one thousandth. In fact, all the planets together are negligible in mass compared with the Sun. Other stars in the Galaxy have

This galaxy has a spiral structure typical of our own and many other galaxies.

masses from one tenth to fifty times the mass of the Sun. In fact, the Sun is only an average-sized star. Our Sun is in the outer part of the Milky Way Galaxy, which has a disk shape with a spiral structure. We obviously cannot take a photograph of our Galaxy until we can learn how to send a probe outside it, but we can take photographs of other galaxies and deduce that our Galaxy probably looks similar. This is supported by the locations of the stars that we do see within our Galaxy. When you look up at the night sky, all the stars that you see with your naked eye are in the Milky Way Galaxy. The ones that are near us can be seen as separate entities, whereas ones that are farther away and closer to the center of the Galaxy form this hazy, luminescent swath we call the Milky Way.

In addition to the stars that make up the disk of our Galaxy, other matter floats around throughout it, in particular, clouds of gas and clumps of dust. Within these clouds of gas new stars are formed. The dust in the Galaxy, made in the outer parts of stars and also by the explosions of stars, obscures parts of the Galaxy from us. When you look up at the night sky, away from city lights, you will notice that the Milky Way is broken by dark spots. Those dark spots are actually clouds of this interstellar dust.

In addition to our Galaxy, we also know that there are other galaxies. One galaxy that you can see with the naked eye is called the Andromeda Nebula (a nebula is a fuzzy patch in the sky). The Andromeda Nebula is a galaxy like our own, with about 100 billion stars, but at a distance of more than a million light-years. In contrast, the size of our Galaxy, as calculated from our position to the center, is only about 30,000 light-years. There are about 100 billion galaxies in our knowable universe. "Knowable" means close enough that light could have traveled to us during the existence of the universe, that is, in about 15 billion years.

The universe is incredibly large. And the story of the structure of objects in it does not end with galaxies. Galaxies, instead of just being separate entities, seem to be grouped together into clusters. For example, the Virgo cluster is a huge group of about 1,000 galaxies in the direction of the constellation Virgo (although much farther away than the stars that form the constellation). We appear to be on the edge of it. Recent observations indicate that the Virgo cluster may in fact be part of an even larger entity. Not only are we not at the center of the solar system, which is not at the center of the Galaxy, but we are not even at the center of our local supercluster. In fact, as we will argue later, there is no center to the universe: all points are roughly equivalent. The clusters of galaxies seem to be arranged not in a completely random way, but into patterns that look like filaments, sheets, and bubbles, among other things. That their arrangement is not random was discovered only in the past few years; this fact has very profound implications for how galaxies and clusters of galaxies first came into being. In Chapter 6 we will see that the distribution of galaxies and clusters might even be related to the fundamental forces that hold quarks and atoms together.

Let us now examine more thoroughly the modern views of matter and interactions. Then we will come back to the study of astronomical objects, and see how their origin and distribution could possibly be related to the nature of fundamental particles and forces.

2 THE ORIGINS OF MODERN PHYSICS

In this computer-generated image, the colored surface represents the spatial curvature of the space-time surrounding a black hole, a region of space with a gravitational field so strong that light cannot escape. Einstein's realization that gravity is the curvature of space-time was one of the major developments of twentieth-century physics.

Although Copernicus and Kepler were responsible for changing our view of the solar system, Galileo Galilei (1564–1642) is generally credited with initiating the modern style of scientific research. His emphasis was not on a philosophical "why," but rather on "how"—on fitting his observations to a mathematical behavior. For example, he was curious about how a falling body's velocity changes as it falls. But he realized that his instruments were too crude to accurately measure the velocity of objects dropped from some height, say from the nearby Leaning Tower of Pisa. However, he reasoned that an object rolling down an inclined plane accelerates like a freely falling body, but with the effect of gravity greatly reduced, depending on the angle of inclination. If the plane were tilted only slightly, the speed would not be excessive and would be easier to measure.

Since stopwatches had not yet been invented, Galileo had to create a time-measuring device himself that could accurately measure the very short intervals from the moment the ball was released to the moment it reached a mark, say, s units down the plane. Galileo's musical talent inspired one such form of a clock. A popular marching tune had a beat, say, every half second. And he, like anyone with a musical ear, could detect an error of a sixty-fourth note, that is, about one sixty-fourth of a second. Galileo put a series of tightly drawn strings across the path of the descending ball. He slid them up and down until the ball's movement past the strings plucked out a rhythm exactly matching the beat of the march. He then measured the distances between strings. Galileo found that the distances between strings, representing equal time intervals, increased geometrically down the plane; that is, the second distance was

Galileo stands at the center of this fresco by Giuseppe Bezzuoli, explaining the uniform acceleration of a sphere rolling down an inclined plane.

four times the first, and the third interval was nine times the first. Galileo stated the results of the experiment in the following mathematical form

$$s = At^2$$

That is, the distance s that a falling body covers is equal to a constant A times the square of the time t it takes to cover the distance. Galileo was keenly aware that the conditions under which he conducted his experiments were far from ideal—friction of the plane, the impediment of the strings, and so forth, affected his results. He took these factors into account when constructing an experiment and analyzing its results—an intuition unique for his time. For example, his demonstration that balls will roll indefinitely on a horizontal plane came from noting how much longer a ball will roll if the plane is more highly polished. This intuitive jump to what we now call Newton's first law of motion was brilliant. He taught his reasoning to his students by the argument of continuity: if the plane is tilted up, the ball, while rolling uphill, will go more and more

Galileo was able to time the rate of fall of a ball under the force of gravity by listening for the click the ball made as it crossed the wires stretched across this inclined plane.

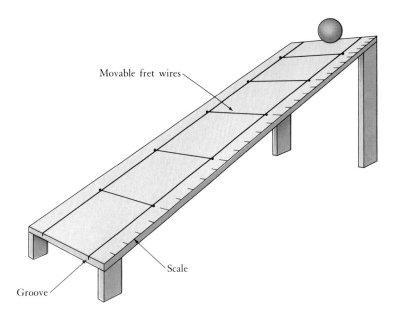

Movable fret wires

Groove

Scale

slowly; if it is tilted down, the ball will go faster and faster; ergo, if the plane is perfect and horizontal, the ball will *neither slow down nor speed up* but continue forever! Thus Galileo understood that *forces* are not necessary for motion, but only for changes in motion.

THE IMPORTANCE OF INSTRUMENTS

Galileo thus transformed science into an experimentally based discipline, ushering in a new breed, the craftsmen and instrument makers—people with magic hands and the ability to visualize how things work. Today, with programmable machine tools and robotic welders, it might seem that these skills are obsolete, but not so. The art of the instrumentalist (engineer) became essential in the seventeenth century, and remains vital to the progress of science in the twentieth.

Telescopes and microscopes, in their most general sense, are the key instruments of our story. Providing a window on two quite different worlds, they share a mechanistic and a philosophical similarity, and both were probably invented in several places in Europe around 1600. Simple lenses had been known for centuries, and the telescope probably resulted from the chance discovery that two spaced lenses, one convex and one concave, would render distant objects "wondrously magnified." It is said

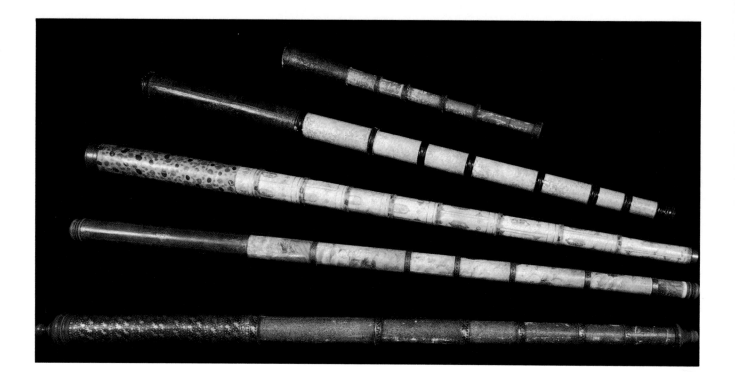

Telescopes from the seventeenth century.

that the telescope was promptly applied as a military device. But Galileo was probably the first to use it for scientific research.

Tycho Brahe, Newton, Huygens, Hooke, Boyle, d'Alembert, and many other famous scientists were skilled instrument makers, but progress in science also depended on an army of lesser-known craftsmen who designed methods for polishing lenses and mirrors; who manufactured optical sights, engraved scales, and quadrants; and who invented such devices as telescopes and microscopes, thermometers, barometers, and pumps.

In the early decades of the experimental age, astronomical surveying was a major activity, motivated by the desire to improve the accuracy in measuring the position of planets and stars, to discover the shape of the Earth, and to make geographic maps more accurate. Later, and more vigorously in the eighteenth century, the pace of research in chemistry and electricity stimulated emphasis on chemical and electrical apparatus.

The giant of the seventeenth century was Isaac Newton, whose laws of motion and theory of gravity placed the work of Galileo and Kepler on a firm mathematical basis and vastly extended the science of mechanics and optics. His work will be referred to in many places in what follows.

THE ELECTRICIANS: COULOMB, FARADAY, MAXWELL, AND HERTZ

The work in the fields of electricity and magnetism that would eventually result in our modern concepts of particle physics and cosmology was the work of many but was led by one Frenchman, two English physicists, and a German.

Charles Coulomb (1736–1806) turned to science late in life. Working in Paris, he invented an extremely sensitive device, called a torsion balance, for measuring the effects of very weak forces. This device consisted of a long quartz or silk fiber suspended from the ceiling and supporting a light, horizontal rod, balanced at its center. Two small pith balls at the ends of the rod were precisely balanced. A force on one ball would cause the rod to turn, twisting the silk fiber. A tiny mirror fastened to the fiber reflected a beam from a strong light source onto a distant screen.

Using this system Coulomb was able to show that the force exerted by one electrically charged ball on another varied inversely as the square of the distance between them (that is, one-fourth as much force at twice the distance, one-ninth at three times the distance, and so on. This is called the inverse-square law.) He was also able to show that the force depended on the product of the quantity of charge on the balls. In addition, he established the attraction of unlike charges (plus and minus) and the repulsion of like charges (plus–plus or minus–minus). But, how are these forces transmitted?

Enter Michael Faraday (1791–1867), a gifted experimenter and an intuitive genius. Trained to be a bookbinder, his education did not include much mathematics, but his intuition was fed by graphic visualizations. Faraday visualized the means by which a charge in one place can exert a force on another charge elsewhere. His concept was that the existence of a charge alters space everywhere. That is, an electric charge creates an electric "field" everywhere. He measured the properties of this field by placing a charge at a position and noting its movement. He concluded that if his test particle experiences a force, there exists a field. The field is presumably established by charges located elsewhere. Faraday imagined lines of force emanating from the first charge—the source of the electric field. The lines gave the direction of the field, and the number of lines passing through a small area perpendicular to the lines measured the field's strength.

Although the field concept was originally thought to be a mere convenience, a way to visualize how forces can act between distant objects, the modern view of forces gives the field a much more substantial role.

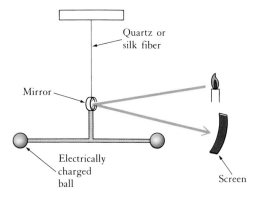

Top: Coulomb's torsion balance. *Bottom:* A light beam reflected from a suspended mirror casts a spot on a screen, which can then show the smallest angle of twist of a long silk fiber.

Quartz or silk fiber

Mirror

Electrically charged ball

Screen

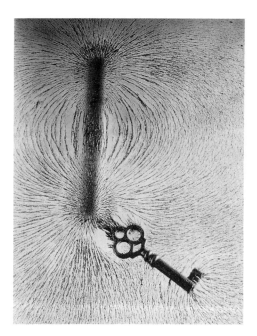

Magnetic lines of force are visualized here by iron filings sprinkled over a bar magnet and iron key. The magnet generates the lines of force, which are distorted by the key.

Soon, others applied the idea to magnets, and almost everyone remembers how iron filings can render lines of magnetic force visible. By 1820, it was known that stationary charges produce electric fields (which exert forces on other charges) and that moving charges, that is, electric currents, produce magnetic fields. These magnetic fields can exert forces on magnets and also on other moving charges. Faraday was struck by this idea that electricity in motion creates magnetism. His intuitive sense cried out for symmetry: Could magnetism in motion create electrical fields? He tried to see if currents flowing in one wire would generate currents in another, nearby wire; they did not. But in one of his many variations of this experiment, he found that at the instant when he connected a coil of wire to his battery, a momentary tiny current surged in a nearby second coil. Further research showed him that the key to producing electric current was in causing a magnetic field to *change*. That is, the current in the second coil flowed only when the current in the first coil was changing. (Today, Faraday's experiment is used on a grand scale. Giant coils are turned in a powerful magnetic field, say, by the mechanical energy of a waterfall, and electric currents, generated in the coil, are sent to light distant cities.) Faraday thus unified what had been considered the separate phenomena of electricity and magnetism, showing that they are aspects of a single, larger concept: *electromagnetism*.

James Clerk Maxwell (1831–1879) undertook to put the known experimental facts of electricity and magnetism, including the pictorial ideas of Faraday, into a compact, mathematical form. His electromagnetic theory would lead to some of the major experiments in this century. Maxwell translated five experimental results into equations:

1. *Coulomb's law and the idea of electric fields.* The connection between electric field and charge is contained in a mathematical constant ϵ, which can be given a numerical value by electrical experiments and a definition of units.
2. *Wires carrying currents produce forces on other wires carrying currents.* This can be visualized by saying that currents produce magnetic fields (Ampere's law). Again, the connection between current and magnetic field involves a constant μ, also given a numerical value by measurement and choice of units.
3. *No magnetic charges exist* (i.e., there are no free poles).
4. *Changing magnetic fields produce electric fields* (Faraday's law).
5. *Electric charge cannot be created or destroyed.* Positive and negative charges can be separated from a neutral system, and will, when recombined, form a perfect neutral system again.

Maxwell wrote each of these experimental laws in a form of mathematics known as differential equations. These equations precipitated a

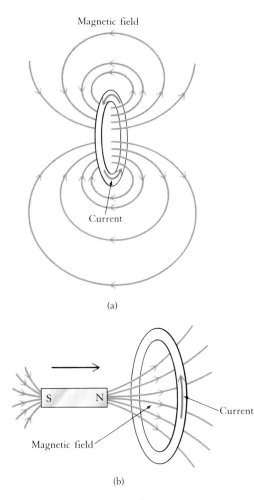

(a) A current moving in a loop generates magnetic lines of force. (b) A moving magnet changes the number of lines of force threading through a loop of wire. This change generates an electrical current in the loop.

crisis: the five experimental laws taken together demonstrated a mathematical inconsistency. The resolution of this crisis was perhaps the most revolutionary scientific event of the nineteenth century.

Maxwell knew that because the laws were based on experimental fact, they had to be correct. But if they were correct, how could they be inconsistent? Maxwell pondered and examined the Faraday pictures. He generalized Faraday's lines of force into stresses and strains in a medium that would fill all space. He called this medium *aether*. It carried the electric and magnetic fields, and seemed both mathematically and intuitively helpful. Maxwell finally noted a lack of symmetry, somewhat akin to Faraday's insight. Maxwell saw that there are *two* sources of electric fields: (1) charges, and (2) changing magnetic fields, but he wondered why there should be only one source of magnetic field—currents. Could changing electric fields also produce magnetic fields? If so, why were these not seen?

Maxwell added this possibility to his mathematical equations describing the observed laws, and the contradictions vanished! What his new equations described was a new view of physical reality. According to his equations, an increasing magnetic field creates an electric field. This electric field grows, and, by its changing, produces a second magnetic field, which hastens the collapse of the first. This collapsing magnetic field in turn generates a new electric field in the opposite direction. Interacting fields thus take on a life of their own, and the equations indicated that once they have started, the waxing and waning fields propagate through space.

Maxwell's mathematics yielded a formula for the velocity of propagation of these electromagnetic waves. The velocity equation contained constants, ϵ and μ, that had been measured in table-top experiments involving charges and currents, wires and meters. When Maxwell plugged these numbers into his velocity equation, he found that the propagating velocity was equal to 3×10^8 meters per second. But 3×10^8 meters per second was also the speed of light measured by means of optical experiments. Maxwell concluded that: ". . . we can hardly avoid the inference *that light consists in the transverse undulations of the same medium which is the cause of electrical and magnetic phenomena.*"

Light was thus identified as an electromagnetic wave: Maxwell had deepened the connection of electricity and magnetism, established the mathematical language of this force, and unified light and electromagnetism. Electricity, magnetism, and optics were all diverse manifestations of a single set of equations. Maxwell's electromagnetic theory, published in 1867, was a triumph of synthesis, rivaling Newton's theory of gravity, and its consequences were equally profound. According to Maxwell, all electromagnetic waves, including light, are generated by oscillating electric charges, and these waves differ from one another *only* in their frequency of oscillation. This theory also predicted the existence of previ-

Frequency, oscillations
per second

Wavelength

10^{22}	
10^{21}	Gamma rays
10^{20}	
10^{19}	X-rays
10^{18}	-1 Å
10^{17}	-1 nm
10^{16}	Ultraviolet
10^{15}	
10^{14}	Visible light -1 μ
10^{13}	Infrared
10^{12}	
10^{11}	
10^{10}	Microwaves -1 cm
10^{9}	
10^{8}	TV, FM -1 m
10^{7}	
10^{6}	Radio
10^{5}	-1 km
10^{4}	Long wave
10^{3}	

The electromagnetic spectrum. The quality and properties of the radiation depend on the wavelength, which is simply related to the frequency of oscillation by the equation: frequency × wavelength = the speed of light.

ously unknown waves, including what we now call radio waves. Could such waves be generated by means of charges and currents in a laboratory? If they could, Maxwell's theory would be verified.

To test these ideas, Heinrich Hertz (1857–1894) began a series of experiments that culminated in 1888 in a complete verification of Maxwell's theory. Hertz found a way of generating oscillating charges, and found that these radiated electromagnetic waves that traveled with the velocity of light and obeyed the laws of optics. Hertz designed an induction coil, a kind of transformer that could deliver very high voltages. He placed this voltage between two metal spheres; when the spheres were brought closer together, a spark crossed the gap. The spark was a very rapid oscillation of electric current in the air between the balls. This was the source of Hertz's electromagnetic radiation. Ingenious detection devices enabled him to measure the frequency, wavelength, and velocity of the radiation. All was as Maxwell had predicted.

In this way it was proved that the universe we live in is filled with a sea of electromagnetic radiation, almost all of which is invisible to the human eye. Only during the last century, since Hertz's experiments, have we learned how to routinely generate and detect such radiation (radio, television, radar, etc.). Progressing from Coulomb to Faraday to Maxwell to Hertz, we see the interplay of theory and experiment on which scientific progress depends.

SPECTROSCOPY

Visible light is physically identical to all other electromagnetic radiation. It is visible to us because our eyes evolved to detect this narrow band of radiation out of the entire electromagnetic spectrum; this band is the dominant radiation from our Sun.

Philosophers and early scientists had speculated avidly about the nature of light. Our modern understanding of light begins with Isaac Newton's prism experiment, in which he demonstrated that when white light from the Sun passes through a prism, it is spread out into the now-familiar rainbow spectrum from red to violet. Newton spent a great deal of effort to prove that the colors were not introduced by the prism but in fact were actually the constituents of white light. It was later proved that every shade of color corresponds to a unique interval of frequencies or wavelengths.

In the eighteenth and nineteenth centuries, the prism used to separate light was embellished by slits and telescopic lenses and thus became a more precise tool for examining light from all sources. Fraunhofer used this spectroscope to discover that the spectrum from the Sun was broken

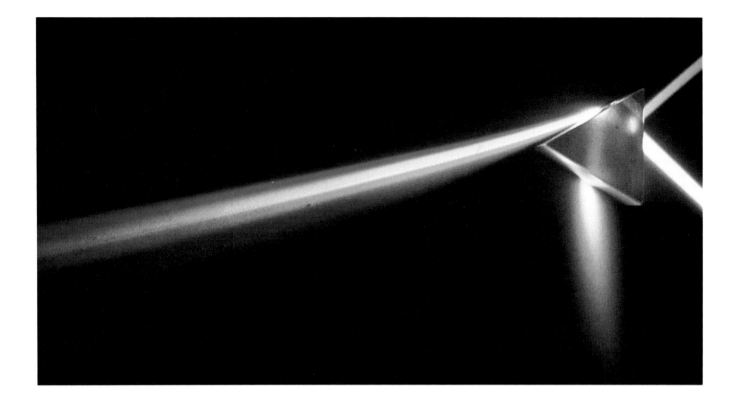

White light decomposes into its component colors as it passes through a glass prism.

by a series of dark lines, whose wavelengths he carefully mapped. In contrast, the laboratory-generated light from heated gases, metals, and salts showed a series of narrow, bright, colored lines on a dark background. The locations of these lines on a wavelength scale were characteristic of each chemical element being heated, and the idea of using these spectra as "fingerprints" to identify the elements being observed gave rise to a veritable industry of careful mapping of spectral lines.

It was also discovered that any element, if heated to a high enough (incandescent) temperature, will produce white light—a complete spectrum of all colors, unbroken by any dark lines. Next came the breakthrough: Incandescent light with an unbroken spectrum was passed through a screen of a chosen chemical element that is at a *lower* temperature. The resulting spectrum had *dark* lines, just like those in the solar spectrum, at precisely the frequencies where that particular chemical element produces its unique bright lines when heated. That is, each element emits and absorbs light only at certain fixed frequencies unique to itself.

The dark Fraunhofer lines in the solar spectrum are the result of chemical elements in the Sun's outer layer absorbing their characteristic

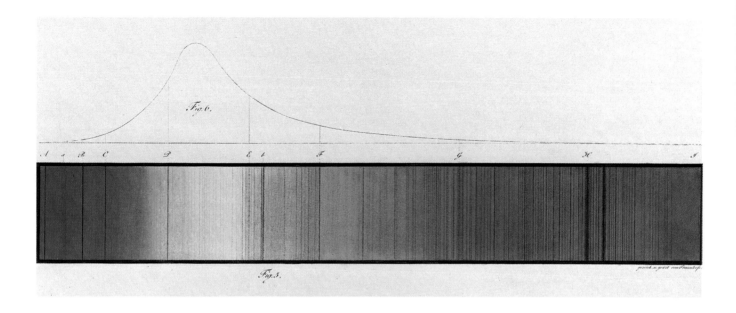

Dark lines observed in the solar spectrum by Fraunhofer. Elements in the cooler surface regions absorb specific wavelengths of light from a purely continuous spectrum of light emerging from the hotter core. The smooth curve above indicates the relative strengths of the different colors (wavelengths) in the continuous spectrum.

frequencies emitted as part of a continuous spectrum produced by the much hotter interior. There were still doubts: In 1878, lines were seen in the Sun that did not match any known spectrum. From them, astronomers predicted the existence of a new element named, in honor of the Sun, helium. In 1895, terrestrial helium was discovered.

Just as Newton's universal theory of gravitation proved that the same laws apply on the surface of the Earth and to the orbits of the planets, so spectroscopy proved that the same chemical elements exist on Earth and in the Sun—and, as we now know, in all the rest of the universe. Physicists suspected that this very selective emitting and absorbing of light that produces spectral lines resulted from characteristics of the "electrical thing" that vibrated within atoms. They began to look for it.

THE ELECTRON: THE FIRST ELEMENTARY PARTICLE

Physicists and chemists knew that Faraday had performed experiments that strongly implied the existence of "units of electric charge." In 1891, Johnstone Stoney, convinced that this charge must be carried by a particle, named it the *electron*.

J. J. Thomson in his Cavendish laboratory with the apparatus he used to find the mass of the electron.

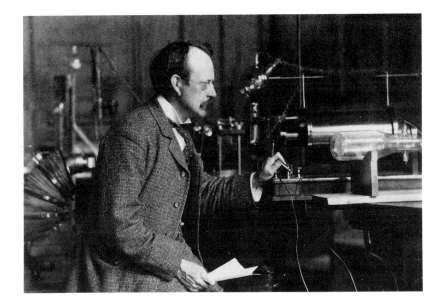

The crucial research on the electron was carried out in Cambridge, England, at the famous Cavendish Laboratory, where J. J. Thomson (1856–1940) was studying electric discharges in gases.

Electric discharge is repressed by the insulating qualities of air at ordinary pressure; so, in order to study such discharges, physicists were pumping air out of glass tubes, to create partial vacuums in which such discharges take place much more easily. It had long been known that if electric voltages were applied to a rarefied gas, usually by sealing wires into two ends of the glass tube, spectacular glows were seen. The systematic studies of this discharge depended on improved pumps, which, by about 1850, could reduce the pressure to one ten-thousandth of ordinary atmospheric pressure. The wire inserts, called *electrodes*, could be of various materials, although platinum was most convenient for sealing into glass. The electrode connected to the negative source of electricity was called the *cathode*, the positive terminal the *anode*, terms devised by Faraday.

Before Thomson's research, it was known that the electric discharge, often indicated by a glow in the residual gas, emanated from the cathode, and could be observed to strike the glass around the anode, causing the glass to light up. Thomson's experiments established that the cathode rays were streams of particles, each carrying a negative electric charge. He was able to prove that these particles also had a mass.

His method was to apply external electric fields and magnetic fields to the cathode rays, and measure their deflections. The fact that the rays

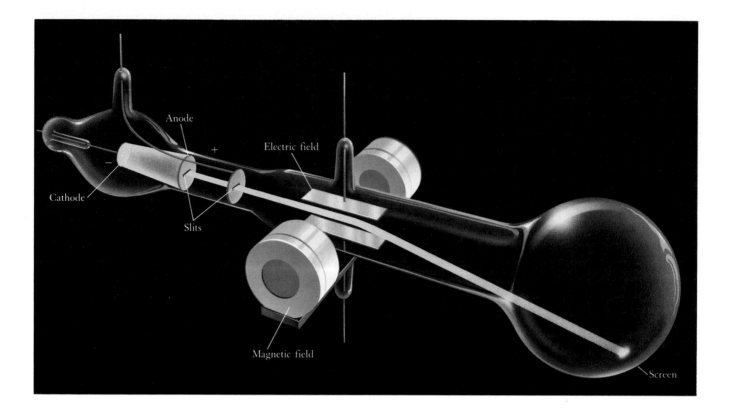

The cathode ray tube of J. J. Thomson. A heated wire at the cathode emits free negative electrons, which are accelerated toward the positively charged slits. A narrow stream continues out of the slits toward the plates. The field of the plates can deflect the stream, which may move up or down depending on the polarity of the field.

were deflected established that they carried charge. Newton's laws of mechanics and Coulomb's laws of electric forces determine the deflection if the electric charge, the mass, and the velocity of the rays are known. By carefully measuring *both* electric deflections and magnetic deflections, Thomson was able to calculate both the velocity and the ratio of the charge *e* to the mass *m* of the particles that composed the cathode rays. The velocity of the rays was found to be proportional to the strength of the battery that established the electric field. Nothing profound here. However, the ratio of charge to mass *was* a profound result. Thomson obtained:

$$e/m = 1.8 \times 10^{-11} \text{ coulombs/kg}$$

He also showed that the value of *e/m* was independent of the gas (he tried many gases), independent of the cathode material, and independent of the voltages and the fields applied. He concluded that this value had to be a characteristic of the particles that make up all matter.

I	II	III											IV	V	VI	VIII	
H 1																	He 2
Li 3	Be 4	B 5											C 6	N 7	O 8	F 9	Ne 10
Na 11	Mg 12	Al 13											Si 14	P 15	S 16	Cl 17	A 18
K 19	Ca 20	Sc 21	Ti 22	V 23	Cr 24	Mn 25	Fe 26	Co 27	Ni 28	Cu 29	Zn 30	Ga 31	Ge 32	As 33	Se 34	Br 35	Kr 36
Rb 37	Sr 38	Y 39	Zr 40	Nb 41	Mo 42	Tc 43	Ru 44	Rh 45	Pd 46	Ag 47	Cd 48	In 49	Sn 50	Sb 51	Te 52	I 53	Xe 54
Cs 55	Ba 56	Lu 57–71	Hf 72	Ta 73	W 74	Re 75	Os 76	Ir 77	Pt 78	Au 79	Hg 80	Tl 81	Pb 82	Bi 83	Po 84	At 85	Rn 86
Fr 87	Ra 88	Lr 89–103															

Lanthanide series (rare earths)	La 57	Ce 58	Pr 59	Nd 60	Pm 61	Sm 62	Eu 63	Gd 64	Tb 65	Dy 66	Ho 67	Er 68	Tm 69	Yb 70	Lu 71
Actinide series	Ac 89	Th 90	Pn 91	U 92	Np 93	Pu 94	Am 95	Cm 96	Bk 97	Cf 98	Es 99	Fm 100	Md 101	No 102	Lw 103

The periodic chart of the chemical elements.

Thomson's discovery was published in 1897. He emphasized that the value of e/m was very large compared with that measured for the hydrogen atom by means of electrolysis. The cathode-ray result for e/m was almost a thousand times larger than that for hydrogen. In 1898 and 1899, Thomson succeeded in measuring the electric charge e separately, using two quite independent techniques. The results were clear and definitive. Electricity is carried by a particle whose mass is 1/2,000 of the mass of a hydrogen atom. Furthermore, Henrick Lorentz had arrived at extremely similar values by analyzing some completely independent data on the emission of light from atoms placed in magnetic fields.

Atoms contain electrons, and light is emitted by atoms; in fact all the optical properties of matter are somehow generated by the vibrations of these electrons. What was needed next was a model of how the electrical pieces fitted together to construct the atoms of the periodic table.

A periodic arrangement of the elements into an ordered table had been formulated between 1869 and 1871 by Dmitri Mendeleyev, a Russian chemist. He proposed that when the known chemical elements are arranged in sequence according to their increasing atomic weights, the resulting table exhibits a periodicity of the chemical properties of the elements; that is, elements with similar characteristics fall into vertical columns in the table. Gaps existed in Mendeleyev's table because many elements had not yet been discovered. But because of the table's periodicity, he was able to predict accurately the characteristics of the undiscovered elements. What was not known in Mendeleyev's time was *why* the elements in the rows of the table form groups with 2, 8, 8, 18, 32, and 32 members and why there is a periodicity of properties. A proper theory of how elections are disposed in the atom would provide the answer.

ERNEST RUTHERFORD, NIELS BOHR, AND THE NUCLEAR ATOM

The end of the nineteenth century was a dramatic period in our story. Two hundred years of polishing Newtonian mechanics had enormously deepened the concepts, embellished the mathematics, and widened the applicability of the laws of motion. Maxwell's equations gave physicists a powerful synthesis of all of the experimental data on electric-

The effect of radioactivity on film is shown here by placing a lump of uranium ore on top of a photographic plate (*right*), and inserting a metal key between the uranium and the plate. After four hours, the plate is developed (*left*). The photo is black where the key protected the film from the radioactivity.

ity, magnetism, and optics. Parallel progress was made in the study of gases and in the concept and applicability of energy conservation. These led to the powerful laws of thermodynamics. Scientists were now able to make accurate descriptions of the motion of planets, the behavior of gases, and the propagation of electromagnetic radiation, including how light behaved in space and in materials. In 1894, Albert Michelson, in dedicating the Ryerson Physical Laboratory at the University of Chicago, said that all that remained to be done in physics was filling in values to the sixth decimal place. In fact, the entire structure of physics would be completely revolutionized in the next twenty years!

The first important hint of trouble came in 1896, when Henri Becquerel discovered radioactivity. He had stored photographic plates wrapped in black paper in a drawer together with some uranium salts. When he developed the plates, he saw they had turned black. Perhaps this accident had also happened to others, and perhaps they had thought there was merely something wrong with the plates; but Becquerel, a good scientist, was alert to the possibilities that the uranium was emitting some kind of radiation, capable of penetrating the black paper. Some simple experiments confirmed this idea. This news spread rapidly around the world. Soon it was discovered that this new "radioactivity" was a spontaneous emission of three kinds of radiation. One form, unde-flected by strong magnets, was given the name gamma rays, and is now known to be electromagnetic radiation of short wavelength. A positively charged component, the alpha particles, was later identified as being the

The products of radioactive decay can be analyzed by subjecting them to a magnetic field (represented by the X's), which in the drawing is perpendicular to the plane of the paper. Alpha particles, which are positive, are deflected upward by the magnetic field, whereas negative beta particles are deflected downward and by a much larger amount. The gamma waves are not deflected.

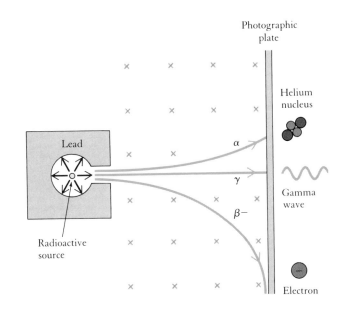

nuclei of helium atoms. These had a mass four times that of hydrogen. The third radiation, the beta particles, were identified later as electrons.

Ernest Rutherford (1871–1937) grew up and was trained in New Zealand. He arrived in Cambridge on a scholarship in time (1897) to witness the discovery of the electron by the Cavendish Laboratory Director J. J. Thomson. Eventually, after a productive period at Toronto and Manchester he returned to the Cavendish in 1919 to reign himself over a tremendous group of younger scientists. In 1908 he was awarded the Nobel Prize for his work on the lifetimes of radioactive atoms.

Rutherford discovered in 1888 that alpha particles are helium atoms minus their electrons. Alpha particles are flung out of radioactive materials at high velocity (about 10,000 kilometers per second). By about 1910, Rutherford recognized that these alpha particles could be used as a probe to poke into atoms to find out how matter is structured. In what became the historical prototype of a scattering experiment, he aimed a stream of alpha particles at gold that had been beaten into a very thin foil. The gold foil was surrounded by a screen coated with zinc sulfide (ZnS). An alpha particle colliding with a ZnS molecule disturbs the binding in the molecule. This situation is unstable, and as the molecule falls back to its more stable state, it emits photons of light. Each alpha particle can plow into hundreds of ZnS molecules, and the resulting flash is visible to the human eye—but just barely. Rutherford and his associates had to sit in a perfectly dark room for several hours to adapt their eyes to the dark to enable them to see the flashes. Both the frequency and location of the flashes had to be recorded—a painfully tedious process. Many repetitions of the experiment were required to give confidence that the results were valid and reproducible.

The result was startling. Although most of the alpha particles passed through the foil with only small deviations from their original direction, some of them *bounced backward.* Rutherford compared his shock to that of someone who fires a 15-inch artillery shell at some tissue paper and sees *it* bounce back. The massive, high-velocity alpha particle must have more than met its match inside the gold atom. Simple calculations, using Coulomb's and Newton's laws, gave an unequivocal result: An enormously powerful electric field must be present inside the gold atom. The known carriers of negative charge, the electrons, were extremely lightweight. The only plausible way such a field could exist is if all the positive charge is concentrated into an extremely small volume.

Rutherford calculated that the positive charge occupies less than 10^{-14} of the volume of the atom! Thus he discovered the *nucleus,* an incredibly tiny, positively charged core, which contains essentially all the mass of the atom. But where are Thomson's electrons? The alpha-scattering research could not tell, since the very low-mass electrons were no barrier for the fast-moving and massive alpha particles. Rutherford concluded that the electrons were spread outside the nucleus in a spherical region which was about 100,000 times larger in radius than the nucleus.

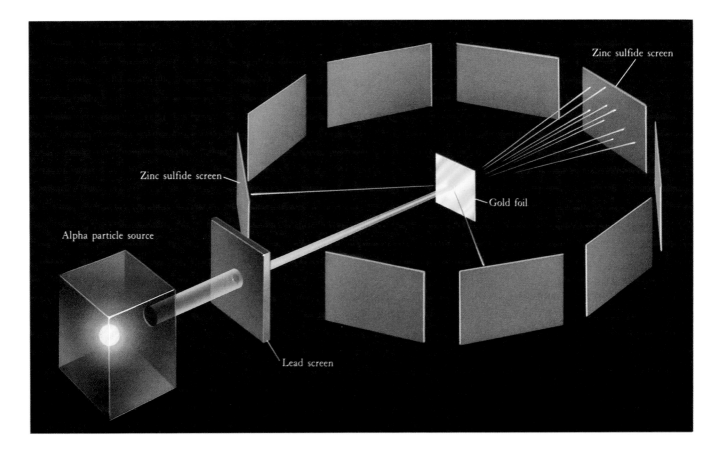

Rutherford's scattering experiment. Most of the alpha particles, collimated into a narrow stream by a lead screen, pass through the thin gold foil with small deflections. A dark-adapted eye can see the tiny flashes as the particles hit the zinc sulfide screen.

Niels Bohr (1885–1962), a young Danish physicist, arrived in Rutherford's laboratory in 1911. At that time the atomic picture was one of a central, tiny positive core carrying a charge just enough to cancel the charge of all the electrons outside it. Now the question arose: How is this atom held together? What stabilizes it? The electron can't stand still; it would fall into the nucleus. The electron could, like the Earth, revolve around the nucleus, with Coulomb's inverse-square law replacing Newton's gravitational force for the Earth–Sun problem . . . or could it? An electron in orbit undergoes circular acceleration, and Maxwell's equations insist that in this situation the electron *must* emit electromagnetic radiation. The electron would then lose energy and spiral into the nucleus. But real electrons don't do that. The model doesn't work!

In other words, the first application of the Newton–Maxwell physics inside an atom failed. It was as if a spaceship had landed on a new planet where the laws of physics were different. Could totally new laws of physics be operating inside the atom? Perhaps this was in Niels Bohr's mind when he proposed a dramatic solution to this crisis. This solution was a complete revolution in thinking—called quantum mechanics.

THE QUANTUM THEORY

Before we describe the basic ideas of the quantum theory, we want to stress two somewhat related general concepts about physics and "the process of progress." We may well ask: How can the classical theories of Newton and Maxwell be wrong? As we have noted, these theories are in accord with an immense amount of experimental data, and in fact are still in current use. They are the basis of all engineering, and are a major ingredient in university curricula in science.

The answer is that these theories are successful when applied to *macroscopic* phenomena, that is, phenomena that depend on the cooperation of extremely large numbers of atoms, so many that the very concept of atoms with internal structure is irrelevant. But within atoms, the theories that evolved from macroscopic observations do not apply; instead, they require drastic modification.

Let's review the basic Newtonian ideas. In the Newtonian view, a particle has a well-defined set of properties: for example, mass, radius, electric charge, location in space and time, velocity, or motion. In Newtonian terms, it was most convenient to use the idea of "momentum," the product of mass and velocity. So, to describe a simple, one-particle system involved stating the position and momentum of the particle at a particular time $t,$ and the forces acting on the particle. If these are known, then the future of the particle would be *completely determined.* This would also apply to a group of particles, say, nine planets and a sun, or a cluster of 1,000 molecules in an evacuated box. In principle (or with the help of a powerful computer), the initial conditions determine the future. In Newton's time, the philosophical implications were devastating. The entire universe, having been set into motion at some cosmic zero time, would develop in a completely deterministic way, its future totally captive to the startup parameters and the forces of nature.

By 1890, cracks—phenomena inexplicable to classical physics—had begun to appear in this Newtonian picture of the universe, and by 1920 some of them had developed into large holes. Let us see what some of these phenomena were.

- Radiation emitted by a black body is incompatible with Maxwell's theory.

- Some experiments had proved conclusively that light consists of electromagnetic waves. Yet other experiments had proved just as conclusively that light consists of discrete particles. How could both statements be true?

- The periodic table of the chemical elements shows that the elements are grouped into families with 2, 8, 8, 18, 32 (and so on) members. But what causes this pattern?

- Why do the chemical elements emit and absorb light at only a few narrow frequencies, which differ for each element?

- Why, under certain conditions, are electrons emitted by a metal surface that is illuminated by a powerful beam of light?

- If electrons orbit around the nucleus inside the atom, why don't they emit radiation in doing so as Maxwell's equations say they should? How *does* the atom work?

There were many other unanswered questions, but these are enough to show that physical scientists were facing an escalating crisis in understanding.

Many physicists, including Planck, Einstein, Bohr, Thompson, and Rutherford, struggled mightily with these problems. Yet the solution, when its final implications became clear, was so radical that it caused grave doubts in the minds of many of the creators—doubts, which for some, like Einstein, were never resolved.

Quantum mechanics solved the puzzles of the strange behavior on the atomic scale by fundamentally altering Newtonian ideas. Let's examine a single object, say, a bead sliding on a wire, whose future positions we would like to know. In Newtonian physics, if our initial measurements of its position and velocity were *exact,* we could predict exactly where the bead would be at any time. However, if our measurements had some error or uncertainty (all measurements do!), we could predict where our bead will be only approximately. By spending more money and time, we could improve the apparatus, reduce the error, and improve the prediction—according to classical physics. Quantum physics states that it is a fundamental law that you cannot continue to improve the precision of observation beyond a certain definite limit.

Werner Heisenberg (1901–1976), a German theoretician, proposed in his uncertainty principle that we can know *either* where a subatomic particle is at a certain moment *or* where it is going, but we cannot know both. The more precisely we know the position of a particle, the less precisely we can know its momentum. Furthermore, Heisenberg's mathematical analysis demonstrated that this situation arises not because of limitations in our experimental techniques (which might be overcome by sufficient expenditures of effort), but because of the nature of the particles themselves. In other words, we cannot achieve the Newtonian ideal of knowing the exact position and momentum of a particle; so the future of any atomic-scale system is not mechanistically determined.

The equations in quantum mechanics that replace Newton's equation have as their solutions only mathematical functions that describe *probabilities*. These functions were first written by Erwin Schrödinger. His equation was able to give a complete analysis of the hydrogen atom, with all the sharp spectral lines accounted for with great accuracy.

The reason Newton's laws and other classical laws of physics work in our ordinary macroscopic world is that we are always dealing with huge numbers of atoms or particles (even a grain of fine sand contains 10^{18} atoms), and statistically the probabilities are near certainties. Although any one particle is not predictable, the behavior of a huge number of particles can be foreseen with great accuracy. Consider the analogy of an insurance company, which can predict the *average* lifetime of male U.S. citizens very reliably, but has a huge uncertainty in its prediction of the lifetime of a single individual citizen. Newtonian mechanics is successful because of the huge numbers of atoms it deals with; the trouble occurs with the single atom, and the solution is the quantum theory.

Quantum effects introduce counterintuitive concepts: unexpected appearances and disappearances, seemingly random fluctuations in the quantities that classical physics held as invariants. Philosophically, what really is taking place is the inherent influence of the observer on what is being observed. It would be ludicrous to try to measure the temperature of a thimbleful of water with a thermometer whose bulb fills the thimble, sloshing water all over the table. A wise experimenter uses an instrument so tiny that it does not disturb the system being observed. How do you do this with atoms? Our available means of observation is not smaller than the atoms, so the atomic observer must allow for inherent influence on the system being observed. It may make us feel better to think of this as the reason for the quantum uncertainties. Quantum mechanics emphasizes, as a way out of innumerable paradoxical situations, that only *measurable* quantities concern us, and that measurements perturb the systems being measured.

The quantum uncertainties blurred the distinction between Newton's hard, massy corpuscle and the probability wave which was needed to describe it in the quantum domain. So, experiments would, on some occasions, see the particle behavior and on others, a wavelike substitute. Symmetrically, light would often be well defined as a corpuscle of electromagnetic energy.

The particle–wave connection was given by the equation

$$E = \frac{h\,c}{\lambda}$$

which related the energy of a light particle (photon) to the wavelength λ of the light now considered as a wave. The constants h and c are the Planck's constant and the velocity of light, respectively. The constant h

first appeared in early attempts by Max Planck (1858–1947) to deal with quantum phenomena. It appears in all quantum effects and its smallness measures the scale at which quantum effects become important. In 1927, the year when quantum theory was rapidly gaining consensus, G. P. Thomson, the son of J. J. Thomson, succeeded in demonstrating that electrons exhibit quintessential wave properties, the characteristic bright and dark fringes. He was able to relate the momentum of electrons to a wavelength:

$$P = h/\lambda$$

Here again, the famous Planck constant appeared. This meant that in the quantum world, electrons can behave like waves. Waves are capable of demonstrating interference effects producing bright and dark fringes. On the level of classical thought this result was totally incomprehensible. A dark fringe means: *no* electrons arrive there. This implied a conspiracy between electrons that arrive on the screen at vastly different times. In wave theory, the dark fringe signals the cancellation of positive and negative crests. But electrons are particles! Quantum theory smugly asserts: no problem! The wave–particle distinction and contradictions cannot be observed *because* of the built-in uncertainty. "If they are not observable, they are irrelevant," is the new catechism of quantum mechanics.

The "quantum" nature of matter on the atomic scale must be clarified. Quantum implies discreteness as opposed to continuousness, it implies abruptness as opposed to smoothness, countable as opposed to uncountable. The electromagnetic field is replaced by light quanta, bundles or packets of electromagnetic energy. Clean, dry sand can be poured and has many of the properties of a homogeneous liquid until we zoom down and look at a few individual grains. We use this metaphor to suggest how the quantum effects can gradually become less important and the classical physics of Maxwell and Newton become increasingly precise as we "go macroscopic." Thus the photon can be called a grain of light.

Since the electromagnetic field carries the electric force, say between two electrons, so the photon must now also assume this task in our theory. Indeed, the quantum theory of electromagnetism expresses the force between two charged particles as carried by an exchange of photons. As we shall see, the photon is only the first of a series of force carriers.

The quantum aspects of nature extend to space itself. Bohr's first step into the Rutherford atom was to assert that only certain orbits of the electron were allowed.

The concept of energy became even more important. The condition of the atom, e.g., a nucleus and its orbiting electron, was more profitably described in terms of its energy states rather than its charge

The circular wave-like orbits are the quantum-allowed orbits of the electron, according to an older version of Bohr's quantum theory of the hydrogen atom. In normal hydrogen, the single electron occupies the lowest ($n = 1$) state, which has a binding energy of 13.6 eV. The energy level diagram below indicates that an electron occupying the $n = 3$ orbit can "fall" down to the $n = 2$ level, emitting a photon whose energy is equal to the energy difference of 19 eV.

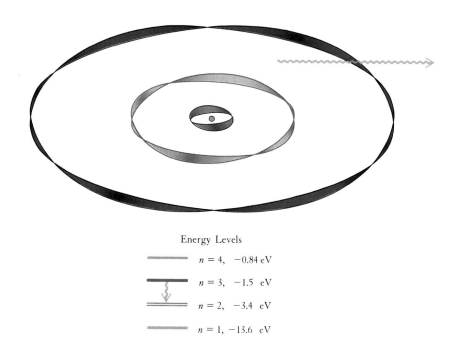

Energy Levels

$n = 4$,	-0.84 eV
$n = 3$,	-1.5 eV
$n = 2$,	-3.4 eV
$n = 1$,	-13.6 eV

configuration. Out of the new rules came the idea that only certain atomic states of well-defined energy were permitted by the laws of nature. These were named, for example, ground state (lowest energy, E_1), first-excited state E_2, second-excited state E_3, and so on. Of course more quantitative designations existed. The mystery of sharp spectral lines was explained by "jumps" between these discrete energy levels, each jump resulting in the emission of a photon of well-defined frequency, energy, and wavelength, for example

$$\frac{hc}{\lambda} = E_2 - E_1$$

The quantum numbers and quantum restrictions were not arbitrarily inserted so as to obtain agreement with the experimental data coming out of atomic physics—we have to emphasize that they all emerged as solutions to the basic mathematical equations (e.g., Schrödinger's equations, whose consequences followed from Heisenberg's uncertainty principles). The sequence involved the major departure of quantum physics from its classical predecessor, the need to deal with probabilities rather than Newtonian certainties.

A vast number of puzzles were solved, a vast number of predictions verified by later experiment. Quantum mechanics worked. The atom was solved. Not only the simplest atoms but even the more complex ones—although precision here required computers working on the basic quantum mechanics equations. All electron orbits were predicted by Schrödinger's equations and, later, by the improved equation of Dirac, which married the special theory of relativity to quantum mechanics. The regularities of the periodic table were understood. Soon after, the mathematics of molecule formation was solved and the chemical bond understood. Quantum mechanics explained electrical conductivity in metals and in the 1950s, the 40-year-old phenomenon of superconductivity was understood. Semiconductors were subjected to quantum mechanics and the invention of the transistor was a direct result. A new discipline arose out of the ashes of classical theory: condensed-matter physics.

Like a benevolent plague, quantum mechanics infected all aspects of science dealing with atoms and molecules. Chemistry was immediately converted and in a decade or so, the subject of molecular biology began its revolution. Not too bad for a theory that starts out by stating fundamental uncertainties.

SPECIAL RELATIVITY

The other revolutionary innovation that dealt with the cracks in the Newtonian picture was Albert Einstein's theory of relativity. The Newtonian concepts of space and time were not very different from those of the ancient Greek philosophers. Space was still an empty receptacle for holding matter; time was thought to "flow" the same at all times and places, always at the same "speed" for all observers. In the theory of relativity, Einstein (1879–1955) unified space and time into a more general concept, space-time, and showed that our experience of location and duration depends on our own state of motion.

Maxwell's equations were the major triumph of nineteenth-century physics. They show that the previously separate concepts of electricity and magnetism were intimately related, and could be unified into one set of equations, and they gave a very precise description of light. Maxwell's equations were shown over and over again by experiment to describe the electromagnetic field in nature, but they implied that the speed of light is always the same whether you are moving or standing still relative to the source of the light. The constancy of the velocity is built into the laws of nature. This is the essential starting point of special relativity.

The special theory of relativity is based on the constant velocity of light. (a) An observer of slowly moving objects finds that the thrower's velocity (the bus) must be added to the ball's velocity to obtain the velocity of the ball relative to himself. He and the thrower would disagree on the ball's velocity. (b) Were the object moving at the light velocity, both observers would perceive the object to have the same velocity, no matter how fast the bus is moving. The result *is* counterintuitive.

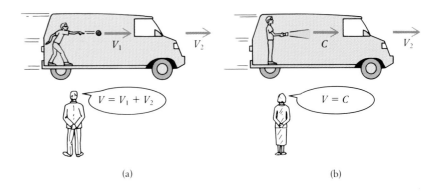

(a) (b)

Let us put this in terms of normal velocities and frames of reference. If you are riding in a bus and throwing a ball back and forth between the front seat and the back, the relative velocity of the ball to the bus may be, say, 10 miles per hour. But for somebody standing on the street, the velocity of the ball would be the velocity of the bus plus the 10 miles per hour that the ball was thrown. If our bus is moving at 65 miles per hour, the ball would be moving at 75 miles per hour as it went by the person standing on the sidewalk. These are very real velocity differences, because if the ball, thrown from back to front, left the bus and hit somebody standing on the sidewalk, it would be moving at 75 miles per hour, the speed of a hard-pitched ball, not a gentle toss of 10 miles per hour.

However, Maxwell's equations implied that light does not operate in this way. The speed of light emerging from the headlights of the bus would be the same whether measured (somehow) by the bus driver or by someone standing on the ground watching the bus go by. That is, light always propagates at the same speed (186,000 miles per second, or 3×10^5 kilometers per second). This problem of light in different frames of reference was a major stumbling block for physics at the end of the nineteenth century. As early as 1887, two American physicists, Albert Michelson and Edward Morley, working in Cleveland, Ohio, demonstrated that the propagation of light on the Earth's surface was apparently independent of the motion of the Earth through space. Many people tried to understand this result by a variety of ingenious mechanisms. However, Einstein, in his special theory of relativity in 1905, took the radical step of recognizing that the constancy of the velocity of light, independent of an observer (or coordinate system), is the fundamental issue. In particular, he argued that the reason the speed of light is the same in all frames of reference is because that speed is the maximum speed that anything can attain; in other words, it is nature's speed limit.

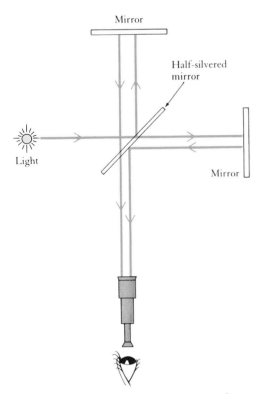

The Michelson–Morley experiment used an instrument called an interferometer which, by measuring the interference fringes arising from the beams proceeding from both mirrors, can give distances up to a precision of less than a wavelength of light (10^{-5} cm). Because the interference pattern did not change as the instrument was rotated, the interferometer showed that the speed of light was the same no matter what direction it was measured from. The velocity of light was observed to be the same whether light moved in the direction of the Earth's orbit or against it.

Nothing can move faster than the speed of light. He thus took the constancy of the speed of light as the fundamental principle, and argued that space and time must change for observers in different states of motion in such a way that no velocity can exceed that of light. Rulers change length, and clocks tick at different rates, for observers in relative motion. If the relative speed is at all close to the speed of light, then these effects should be observable. Indeed, just such assumptions had been made to understand the Michelson–Morley experiment.

In effect, Einstein argued that Maxwell's experimentally verified equations were more reliable than our intuitive notions of space and time. This theory was called *relativity* because it dealt with the relative behavior of systems in different frames of reference. The fact that the speed of light is the same in all reference frames or coordinate systems was consistent with the more general statement that the basic laws of physics are independent of the coordinate system used to make the observations.

With relativity, space and time were no longer fixed, "God-given" quantities, but were shown to be dependent on the frame of reference from which the observation was made. Einstein's theory dealt with how measurable quantities transform between frames of reference moving at fixed relative velocities. This special theory of relativity has been verified again and again in elementary particle experiments, where it is comparatively easy to find objects moving with velocities near the speed of light. For example, one consequence of special relativity is that when particles are moving very fast, approaching the speed of light, time slows down for those particles relative to the observer. Thus, an unstable particle that at rest might decay in a very small fraction of a second, if accelerated to a velocity approaching the speed of light, will be found to live much longer than its lifetime at rest. Stated more carefully, an observer moving along with the particles will measure a certain lifetime. This is the "at-rest" lifetime. An observer in a laboratory in which the particles have a high velocity, say, 0.90 times the velocity of light, will measure a *longer* lifetime—over twice as long. This is often stated by the phrase "moving clocks run slow."

Another consequence of this relativistic transformation is the realization that energy and mass must be related, since they too are affected by moving from one frame of reference to another. This led to the famous relationship

$$E = mc^2$$

which says that mass m and energy E are one and the same thing, just measured in different units, and to correct for those units we must use the maximum velocity of the speed of light c.

GRAVITY, CURVATURE, AND GENERAL RELATIVITY

In the preceding discussion we've seen how special relativity was a natural development from Maxwell's equations. Einstein needed more than ten years to extend his concept of special relativity to the generalized situation of changing velocities (accelerations) and forces. The particular force Einstein treated was the gravitational force. This generalization to gravity is called the general theory of relativity.

The general theory was published in 1916 and is far more complex and difficult mathematically than special relativity. General relativity deals not just with simple transformations between frames of reference with constant velocity, but with frames of reference that are accelerating. In general relativity, Einstein took the concept of interrelated space and time, and moved it one step further by talking about curved space-time. He argued that gravity actually is the curvature of space-time.

The idea of our three-dimensional space being curved is conceptually important. We can, of course, envision curved two-dimensional surfaces in our regular three-dimensional space. For example, the surface of the Earth is a two-dimensional sphere embedded in our apparently flat three-dimensional space. To extend one step further and say that our three-dimensional space is a curved space embedded in a larger four-dimensional space-time is difficult for us to visualize, because we are embedded in that space, and so we see it as being flat. However, there are some tests that we can do to see whether or not our three-dimensional space is indeed curved.

To do this we need to remember how curved geometry is different from flat geometry. For example, if we draw a triangle on a flat, two-dimensional surface, the sum of the angles equals 180°. However, if we draw a triangle on the surface of the Earth, and we make the triangle large enough (for example, going from the north pole to the equator, along the equator, and back up to the north pole) we can end up with triangles that have the sum of their angles far larger than 180°. In fact, each of the angles in our large triangle can be a right (90°) angle, an impossibility in plane geometry. In so doing, our triangle now has a sum of angles of 270° rather than 180°. Clearly, triangles provide us with a way of testing whether space is flat or curved. Of course, to do this for the Earth, we had to use a large enough fraction of the Earth's surface to see major differences. Thus, to see something about the curvature of the universe, we would have to have scales that are an appreciable fraction of the universe itself, and we do not. The universe is so large that no observations we have made to date have been able to tell us directly whether it is curved or flat.

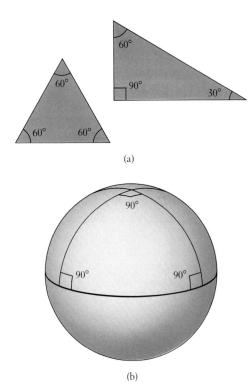

(a)

(b)

Triangles in (a) flat and (b) curved two-dimensional space.

In this computer graphics image, the spatial curvature of the space-time surrounding a black hole is represented by the curvature of the embedding surface. The colors on the surface are related to the relative speed of clocks at those points (red represents the slowest and blue the fastest clocks).

To understand the amount of curvature that is produced by a gravitational field, let us go back to our realization that c is the *limiting velocity*. Thus, a significant curvature would be the curvature of space that prevents even light from escaping from a region. A black hole is a region of space so curved upon itself that light cannot escape. For example, if our Sun were squeezed to a radius of 3 kilometers, it would become a "black hole." The German physicist (and soldier) Karl Schwarzschild was dying in a hospital during World War I when he heard of Einstein's theory. Just before his death Schwarzschild worked out the consequence that if a gravitational field were powerful enough, it would cause space to curve on itself and cut itself off from the rest of the universe. This concept we now call a black hole.

Observations in X-ray astronomy have found objects in space that appear to be black holes. We can find these objects using X-ray astronomy, but not because they directly emit X rays. X rays are electromagnetic radiation, and they too would not escape the black hole. However, when matter falls onto a black hole, it is crushed to such high densities and temperatures that it emits X rays before actually vanishing into the hole. These X rays emitted outside of the black hole can then be detected by X-ray telescopes flying on satellites around the Earth. The most famous example of a black hole is the X-ray source Cygnus X-1, the brightest X-ray source in the constellation Cygnus. This object is thought to be a black hole in orbit around another star. The other star rains debris onto the black hole, and this is the source of X rays. General relativity is important on scales where curvatures approach those of a black hole. If the density of the universe is sufficiently high, then, according to general relativity, the universe would be curved on itself and nothing could escape. It would eventually collapse to near infinite density. As we shall see, an open question in cosmology today is the precise value of the density of the universe.

General relativity has been tested by measurements of the behavior of astronomical objects and of light in strong gravitational fields. These have shown that it really does provide a good description of nature. For example, general relativity accurately predicts how light gets bent as it travels near the Sun, and the small curvature of space near the Sun is sufficient to explain an anomaly in the orbit of the planet Mercury, inexplicable in Newtonian physics. The instruments used to measure the bending of light and the position of Mercury are telescopes, which initially "saw" in optical light but now can also see in other wavelengths.

THE CURRENT CRISES

We've seen in this chapter that there were two major developments that characterized the twentieth-century revolution in physics: quantum mechanics and relativity. Quantum mechanics told us that nature at the very small level did not behave in a classical manner. We also found that at very high velocities, space and time did not behave in the classical way either. When we deal with gravitational fields, space-time is curved rather than flat. Each of these new revolutionary developments has been verified repeatedly as an accurate description of nature in the regimes where they could be tested. At small scales, quantum mechanics describes how particles interact. At large velocities, special relativity describes their interaction. In high gravitational fields, such as that of Cygnus X-1, general relativity provides a very good description. Again, relativity and quantum mechanics do describe how the universe behaves.

Special relativity, quantum mechanics, and electromagnetism were married in the 1940s into a field called quantum electrodynamics. A major goal of modern physics is to try to merge the areas of quantum mechanics and general relativity. The one works very well on the small scale; the other works very well on the large scale. However, as we shall see in Chapter 5, the early universe was hot and dense. Important cosmological distances were of subatomic size. That is, the entire new-born universe operated at sizes that are subject to quantum mechanics. To understand the early universe, we must merge these two revolutionary ideas into one. The problem with this merger is that we don't yet have extensive enough experimental data to guide us as we did when each of the earlier revolutions was worked out.

When we use particle accelerators to look into the very small, we can explore the world of quantum-mechanical systems. Also by accelerating particles to high velocities, we can verify special relativity. By looking at astronomical bodies where gravitational fields are large, we can test the properties of general relativity. But the domains where general-relativity effects turn out to be important are the exact opposite of the kind of objects where quantum mechanics is important. We have a problem.

Modern physics is trying to address this problem by looking at the coupling together of the very large and the very small at the birth of the universe. By trying to use what bits and pieces of clues we have left over from the birth of the universe, we hope to see whether we can piece together how the quantum–gravity interaction must have occurred in order for the universe to look the way it does. This means that physicists must combine all they know about quantum mechanics with what they know about gravity, and then try to extrapolate that combined knowledge back to the early universe. It is the new particle accelerators that may give us hints about these extrapolations, and that have helped make cosmology an experimental science.

3 THE DETECTION OF SUBATOMIC PARTICLES

A bubble chamber photograph of the collision of a high-energy neutrino with the nucleus of a hydrogen atom. The chamber is in a strong magnetic field that curves the path of charged particles into circles. The spirals indicate electrons losing energy as they progress through the liquid hydrogen. Visible as well are the tracks of pions, kaons, protons, and lots of photons converting into electron–antielectron pairs.

Physicists read balances, gauges, thermometers, and barometers; use clocks and rulers; measure spectral lines and other forms of radiant energy; observe the blackening of film—all of which record macroscopic phenomena, but at the end of the nineteenth century these devices began to tell us about microscopic things: electrons, ions, and atoms. In the twentieth century came devices whose sole purpose was to detect events on the atomic scale.

The invention of these devices was spurred by the discovery of natural sources of subatomic particles whose high energy made them easily detectable. The first of these natural sources was radioactivity, discovered by Becquerel in 1896. As mentioned in the previous chapter, Becquerel noticed that certain materials spontaneously emit radiation that can penetrate thick paper to blacken photographic film. The second source of subatomic particles was the cosmic radiation first identified by Victor Hess in 1912. These two sources are composed of individual particles that travel at great velocities. The nuclei of such naturally radioactive elements as uranium throw out alpha and beta particles with very high energies. In cosmic rays, particles with energies thousands to millions of times greater than even alpha and beta particles are rained on our planet from outer space. The 1933 Chicago World's Fair opening ceremonies began with the amplified signal from a particle that started its journey to Chicago millions of years ago.

THE GEIGER COUNTER

Energetic particles can be detected when they collide with atoms, creating reactions that can be amplified to give discernible signals. We have already referred to Rutherford's zinc sulfide screens that detected alpha particles. One of the earliest electrical devices for detecting particles was the famous Geiger counter, invented in Rutherford's laboratory in 1910. The "click" of a Geiger counter is familiar to moviegoers, science-fiction fans, and radiation workers, each click representing the passage of a single atomic-scale particle. How does the Geiger counter work?

Inside the Geiger counter, a thin wire runs down the center of a metal tube and is brought out through glass seals at the end so that the wire is electrically insulated from the tube. With suitable gases in the tube and a large (approximately 1,000 volts) positive voltage on the wire, a single charged particle passing through the tube can easily be detected. As it proceeds through the gas inside the tube, electrons are released from their atoms by the many collisions of the incident particle. The freed electrons head for the wire, colliding with gas atoms and releasing new electrons; and eventually an avalanche of electrons reaches the wire. The resulting current can be amplified and made to "click" a number register. This counter was widely used in cosmic ray research and as an inexpensive and rugged detector of radiation. As we shall see, it was replaced by superior devices in research carried out with accelerators.

The Geiger counter. The passage of a charged particle produces a rush of current in the wire, which can click a register. A Geiger counter is typically 10 inches long and 1 inch in diameter.

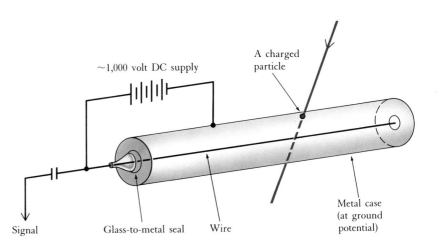

~1,000 volt DC supply

A charged particle

Signal

Glass-to-metal seal

Wire

Metal case (at ground potential)

THE CLOUD CHAMBER

A powerful variant of the zinc sulfide screen is the *Wilson cloud chamber,* invented by C. T. R. Wilson about 1894. In this device the trajectory of a single charged particle is rendered visible and can be photographed. Wilson (1869–1959) was a Scottish scientist who was fascinated by the physics of the clouds, fog, and rain, frequently found in his part of the world. Meteorologists will occasionally assure us that there is 100% relative humidity, which means that there is as much water vapor suspended in the air as the air can hold. Under these conditions we say the air is saturated. How does this water vapor convert to rain?

The process of droplet formation is subtle, but apparently requires some kind of "nucleation center" around which the water vapor can condense. This is normally supplied by dust in the air. In the process of artificial rain-making, stubborn clouds are seeded with crystals to encourage the nucleation of drops from the ample supply of water vapor. Under contrived laboratory conditions in the absence of dust, water vapor will not form drops, even if it is supersaturated. In the cloud chamber the temperature of a container of clean air is suddenly lowered, leaving *more* water in the vapor state than the air can normally support.

Into this unstable situation comes a rapidly moving, electrically charged particle (also known as an incident fast particle). It collides with the air molecules in its path, merely nudging some, without transferring significant energy. But about thirty times in each centimeter of its pas-

An early Wilson cloud chamber. A camera looking down into the cylindrical, gas-filled chamber would photograph droplets of liquid formed on the trail of energetic particles. Droplets could form only during a brief time after the chamber is made sensitive by a sudden expansion of a piston underneath.

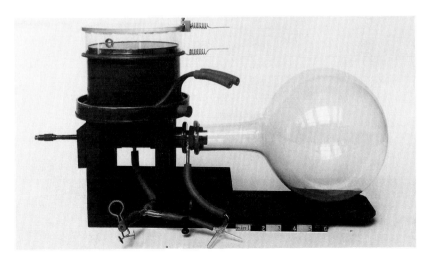

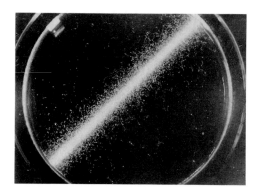

This photograph was taken by Wilson while researching the operation of his cloud chamber. An intense beam of X rays passed through the chamber, ejecting many hundreds of low-energy electrons. Each electron collects only 5 to 10 drops.

sage, collisions are strong enough to remove an electron from the oxygen or nitrogen atoms in its path through the chamber; that is, the incident particle *ionizes* them. The influence on the incident particle is small, a minuscule decrease in its energy, a very tiny deviation from its original direction, but the energetic particle leaves behind it a trail of ionized atoms. Now, there is nothing more appealing to the water vapor, eager to condense, than an electrically charged object, and tiny drops of water form on each ion. These "nucleated" droplets grow rapidly to visible size in a few thousandths of a second, creating a trail somewhat analogous to the vapor trail of a jet plane in the upper atmosphere. If a strong light illuminates the chamber, an observer can see and photograph the path of the particle.

After the discovery of radioactivity, Wilson experimented with his cloud chamber to see if it could detect radiation of different types, but it wasn't until 1910 that he succeeded in seeing the tracks of charged particles. In 1927, Wilson was awarded the Nobel Prize for his invention. The Wilson chamber became the workhorse of nuclear and particle physics until it was replaced in the 1960s by the more effective bubble chamber.

COSMIC RAYS

Exploding stars or supernovas (such as the famous one in February 1987) emit tremendous amounts of energy, much of which is in the form of very high energy particles, some with energies beyond those of accelerator experiments. These particles are known as *cosmic rays*. Thus, even before we had accelerators on Earth, we had the high-energy particles raining on the planet from outer space.

It was first noted that these particles were indeed coming from space in an experiment carried out in August 1912 by Victor Hess. Hess (1883–1963) used a hydrogen-filled balloon with an ionization counter. This device measured the intensity of ionization that results when energetic particles separate normally neutral atoms in a gas into free electrons and a positively charged residue. Unlike the Geiger counter, there is no avalanche; the liberated charges are collected, and the amount of charge is a measure of the intensity of ionization present. Hess noted that there was a certain rate of ionization at ground level. The higher the balloon rose in the Earth's atmosphere, the more the rate of ionization *increased*. Hess did this experiment both during the day and at night, and found no change, thus proving that the source of the ionization was unrelated to the Sun. Whatever was causing this ionization was more intense at higher altitudes than at lower altitudes, so it was not coming from radioactive decays in the Earth. It was instead coming from something outside the Earth and not from the Sun. In this way Hess argued that these charged particles were coming from outer space. It took fifteen years before the extraterrestrial origin of cosmic rays was generally accepted. (In 1936, it was demonstrated that most cosmic-ray particles striking the upper atmosphere of the earth are extremely energetic protons.)

The problem of understanding cosmic radiation was complicated by the fact that particles hitting the Earth would strike atoms of the atmosphere, generating large amounts of secondary radiation, which in turn would interact in the lower atmosphere. What reaches sea level is then quite different from the primary radiation.

THE CONCEPT OF ENERGY

In the few years following Hess's work, James Chadwick would discover an apparent violation of natural law that would eventually lead to the discovery of a new particle, the *neutrino*. The discovery of the neutrino and other subatomic particles through the use of detectors would depend on understanding fundamental principles concerning energy.

Let us carefully define the concept of energy as used by physicists. An object gains energy by increasing its velocity. In the simplest system of a slow-moving "point" object (i.e., an object with extremely small or even zero radius) having a mass m and a velocity v, the *kinetic* energy (energy of motion) is equal to $mv^2/2$. An object can also gain energy by virtue of the forces or fields in which it finds itself. Think of an object at the end of a compressed spring or at the top of a building (see the figure on the following page). When released, the object will gain kinetic energy. Before being released, it has *potential* (stored) energy.

The energy concept is fruitful primarily because of the principle of the conservation of energy: The sum of potential and kinetic energy is constant throughout the motion of an object. Furthermore, real objects, as opposed to point objects, can also have internal energy, for example, a piece of coal, or an electric battery. Energy can be converted from any one of its many forms to another in the total activity of a system, and a careful accounting would reveal that the total energy is exactly the same before, during, and after any activity. For example, an artillery shell in midair just before it explodes would have the kinetic energy of its motion, potential energy due to its location in the gravitational field of the Earth, and chemical energy of the explosives. The unenviable job of verifying the conservation law after the explosion would require measuring the kinetic, potential, and internal energy of each piece of debris, measuring the increased internal (heat) energy of the hot gases that were generated and *their* kinetic energy, and so forth. We now recognize many forms of energy: energy of motion, potential energy in a force field

Energy is continually being converted from one form to another. At the top of the tower, the energy of the ball is all potential energy; at the bottom, after the force of gravity has acted on the ball, its energy is all kinetic.

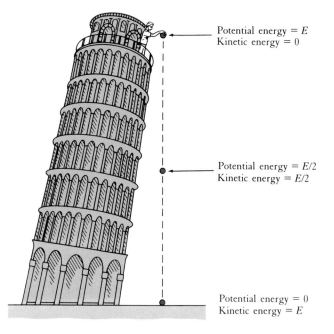

Potential energy = E
Kinetic energy = 0

Potential energy = $E/2$
Kinetic energy = $E/2$

Potential energy = 0
Kinetic energy = E

(e.g., gravity, or that produced by a compressed spring), chemical energy, electrical energy, and so on—all interchangeable, given the appropriate devices.

Untold thousands of experimental tests (probably none included an exploding shell!) have led to complete verification of the law of conservation of energy, which is not a new law of nature, but rather a *consequence* of the laws of nature that seem to govern our universe. Other conservation laws exist, among them the law of conservation of momentum. The *momentum* of an object is defined by its mass times its velocity *mv.* This product is a vector; that is, it has both a size and a direction. The law states that the total momentum, in any given direction, of an isolated system is constant. Conservation laws deduced from observation and measurement are eagerly sought because they are important clues to the structure of nature.

A MEASURING SCALE FOR ENERGY

To discuss the energy of subatomic particles, we need a new unit of measurement. Elementary physics texts use the *joule* as a unit of energy. The joule is conveniently defined as the potential energy of a 1-kilogram mass lifted to a height of 1 meter. It is also the kinetic energy of the mass just before hitting the ground after being dropped. An alternative measure of energy familiar to home owners is the kilowatt hour, the unit by which our electrical bill is calculated. One kilowatt hour is equivalent to 3.6 million joules. You get a lot of joules for about five cents!

On the microscopic scale, however, even one joule is a huge energy; so instead we will use the *electron volt* as our unit; it represents the energy acquired by a single electron when it is "lifted" to 1 volt by a battery. Equivalently, if an electron crosses a gap between two electrodes that are connected to a 1-volt battery, it gains a kinetic energy of 1 electron volt (1 eV). It takes 6×10^{18} eV of energy to equal 1 joule. This enormous ratio is another reminder of the scale of microscopic quantities compared to our own human scale.

We first define the rest energy of an idealized point particle (with no internal energy) as:

$$E_0 = m_0 c^2$$

Here the subscript zero indicates that the mass of the particle is to be measured when the particle is at rest. This is needed because the theory of relativity also implies that the mass of an object *increases* as its velocity increases. The formula is

$$m = \frac{m_0}{\sqrt{1 - (v^2/c^2)}}$$

Clearly, if the velocity of the object is *much* smaller than the velocity of light c, then v^2/c^2 is very small; so the square root is very nearly equal to 1 and m is essentially equal to m_0. However, for microscopic particles, velocities close to the limiting velocity of light can generate a very small number in the denominator of the equation, and a huge increase of m over m_0. (Our present powerful accelerators produce particles whose mass has been increased by 100,000 times.)

It is often convenient to talk about the *total* energy of a particle, and the correct equation is the familiar

$$E = mc^2$$

Mass and energy are so intimately bound by this equation that we often talk about the mass in energy units (eV); that is, we multiply m by the constant c^2. For example, the mass of an electron is more conveniently given in electron volts as 0.511 MeV than in its more conventional value in kilogram units.

ELECTRON VOLT NOTATION

1,000 eV = 1 keV (thousand, 10^3 eV)
1,000 keV = 1 MeV (million, 10^6 eV)
1,000 MeV = 1 GeV (billion, 10^9 eV)
1,000 GeV = 1 TeV (trillion, 10^{12} eV)

MASS OF PARTICLES IN eV

Electron
  m_e = 0.5110041 ± 0.0000008 MeV
Neutron
  m_n = 939.5731 ± 0.0027 MeV
Proton
  m_p = 938.2592 ± 0.0052 MeV

When the velocity v of a particle is much smaller than c, the energy E is equal to the sum of the kinetic energy and the rest energy:

$$E = \tfrac{1}{2} m v^2 + m_0 c^2$$

The temperature of a system measures the average energy of the particles in the system. A temperature of 10,000 degrees on the Kelvin scale is equivalent to an average molecular kinetic energy of 1 eV. Therefore, if an object is at room temperature, the average energy of its molecules is one thirtieth of an electron volt.

CHADWICK, PAULI, AND THE DISCOVERY OF THE NEUTRINO

James Chadwick (1891–1974) was one of Rutherford's most productive students. His first significant work concerned the manner in which beta rays were emitted from radioactive nuclei. His results eventually led (around 1930) to the "invention" of the neutrino.

Chadwick set about measuring the energy of the beta particles by asking: Did they all emerge with the same energy, or did they have a distribution of energies? Many had tried to answer these questions, including the great Rutherford, and had obtained misleading results. Chadwick had won a fellowship to study in Berlin, unfortunately in the year just before the outbreak of World War I! There he set up an experiment in which a magnet was used to measure the energy and momenta of electrons emerging from radium. (Particles are deflected by a magnet, and the amount of deflection is a measure of the momentum.) To detect the electron, Chadwick used an electric-discharge counter, similar to a Geiger counter. All previous research had made use of photographic film to record the arrival of electrons. Chadwick obtained his results shortly before he was interned by the Germans. He verified that the particles emerge with energy anywhere along a continuous spectrum.

The significance of Chadwick's finding was that it seemed to violate the law of the conservation of energy. The reaction is

$$A \rightarrow B + \text{electron}$$

A is the radioactive atom before decay; B is the atom after decay. The electron can have energies that vary from almost zero up to some maximum value. The total energy of A including its rest-mass energy must be equal to the total energy of B plus the total energy of the electron. Now, in each of the disintegrations, A is always the same. It turned out that the energy of B was almost totally independent of the electron energy, which was observed at times to be high, and at times to be low. The energy equation refused to balance! Was an unseen particle also being emitted? The only candidate at that time was the photon, since this electrically neutral object does not cause a Geiger counter to click or leave tracks in a Wilson chamber. This possibility was ruled out by the use of a new technique of measurement in nuclear research, the calorimeter, applied by Charles D. Ellis, who had been a fellow detainee in the German prison camp with Chadwick.

After the war, Ellis joined the Rutherford team. He and William A. Wooster devised a method for capturing all of the energy that could be emitted in the transition $A \rightarrow B$. The trick was to let the emitted energy heat up a large cylinder of lead, thermally insulated and equipped with a very sensitive device that measured small changes in temperature (thermocouple). The result was definitive: Nothing else was detected to come off; the average energy released in the reaction was far less than the difference in energy between A and B. The results were soon repeated on the continent, but neither Ellis nor others commented on what it all meant. By 1929, the problem of the "lost energy" in beta radiation had grown so intense that Niels Bohr suggested that perhaps the law of conservation of energy did not apply in the domain of the nucleus.

The explanation for the "lost energy" had to wait for Wolfgang Pauli (1900–1958). A Swiss theoretical physicist, Pauli made his reputation at the age of 19, when he wrote a review of Einstein's general theory of relativity that is still a model of clarity and incisiveness. Pauli, at age 25, was also the author of the famous Pauli exclusion principle, which first explained the shell structure of electrons, that is, the build-up of the electron orbits as one proceeds through the periodic table.

A neutrino collision photographed in a spark chamber, a detector that produced a series of sparks along the trajectory of the two emerging tracks. Each gap is formed by two one-inch-thick aluminum plates. The track in the center sloping up to the right is a muon. The short, lower track is probably a pion. Since all the neutrinos gave muons, the experiment established the muon's neutrino.

Pauli was to become the critic and even the conscience of physicists because of his fierce devotion to high standards of clarity, but he was also charismatic, and stories that verge on legend rose around him.

Pauli could not accept Bohr's idea of a breakdown of the conservation of energy principle, and, in a "desperate way out," Pauli proposed in 1930 the existence of a new particle that escapes from the reaction without leaving a track and without depositing heat in Ellis' calorimeter. This particle had to be neutral, very penetrating, and of low mass. Pauli thus predicted the *neutrino,* as Enrico Fermi was to name it later. Thus, the beta-decay reaction is actually

$$A \rightarrow B + \text{electron} + \text{neutrino}$$

as Fermi wrote in 1933.

Twenty-five years would pass before the neutrino was detected directly, but long before that indirect evidence, deduced from application of the conservation laws, led to a general acceptance of the neutrino idea.

CHADWICK AND THE NEUTRON

In 1920, Rutherford had already speculated on whether an electron and a proton could combine into a neutral object smaller than the hydrogen atom. Chadwick pursued Rutherford's idea, achieving an answer after twelve years of research. When success finally arrived, the actual discovery took "only a few days of strenuous work."

During the heady days of the worldwide race to understand radioactivity, many new radioactive substances were discovered and used as sources of radiation. In Germany in 1928, alpha particles from the radioactive element polonium were directed at a piece of beryllium metal to study the ensuing reaction. A penetrating radiation, which carried no electric charge, was observed to emerge. In Paris, this radiation was also observed, and attempts were made to interpret it as gamma rays, by then known to be high-energy photons; but these were soon ruled out.

This failed attempt attracted Rutherford's interest, and he pointed it out to Chadwick, who determined to investigate the problem. Chadwick combined polonium and beryllium into a new source of the mysterious new radiation. This he directed at an ionization chamber, successively filled with hydrogen, helium, and nitrogen. In each case, the unknown radiation would occasionally impact a target-gas atom, producing recoils whose energy could be measured. In this way Chadwick was able to calculate that the mass of the new radiation was roughly

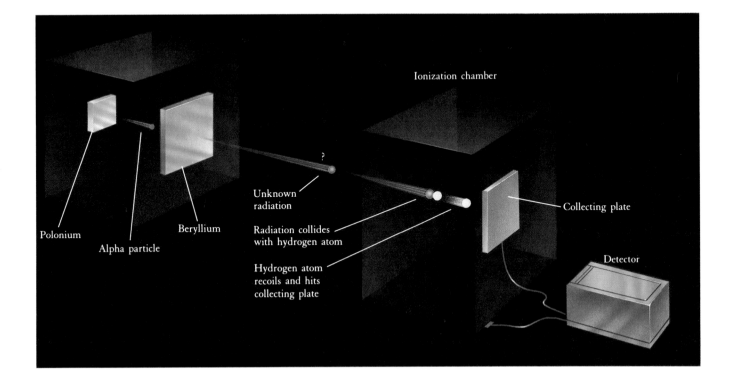

Chadwick's discovery of the neutron. An intense stream of alpha particles strikes a piece of beryllium, producing the mysterious neutral radiation. The neutrons occasionally strike the nucleus of a hydrogen atom in the ionization chamber. Since the colliding objects are of roughly equal weight, the hydrogen recoils from a head-on collision with the velocity of the incident neutron. By moving the collecting plate, Chadwick could ascertain the range of the recoiling hydrogen, and this information could be used to estimate the mass of the neutron.

equal to the mass of the proton. He had discovered the *neutron*. This discovery greatly clarified the structure of atomic nuclei, which were soon considered to be composed of protons and neutrons.

THE STRONG FORCE
AND BINDING ENERGY

The forces between neutrons and protons, neutrons and neutrons, and protons and protons must be very strong and have a very short range; otherwise, the coulombic force pushing the positive charges out would blow the nucleus up. The force holding the nucleus together is appropriately called the *strong force*. Gradually, as a result of many experiments, a quantitative mathematical description of the behavior of this force relative to the distance apart and the state of the colliding particles became known. The pattern of how nuclei are built up from

The binding-energy curve tells all. Since the very heavy ($A \sim 200$) nuclei have less binding energy than medium nuclei, the fission of the heavy elements results in a release of energy and a more stable final configuration. Similarly, light elements release energy by fusion.

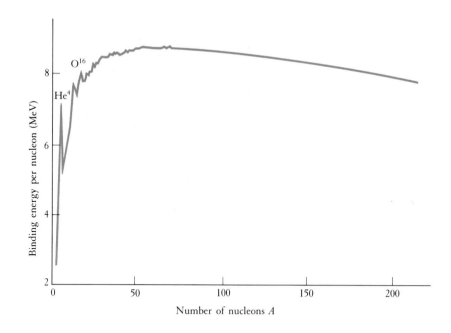

protons and neutrons was also understood when these forces were used to compute the various internal-energy properties of nuclei, crowded with neutrons and protons.

The importance and momentous consequences of the binding-energy curve also became apparent. This curve is shown in the graph on this page. When we deal with complex systems, such as nuclei, atoms, and molecules, the internal energies are determined by the arrangement of the parts and the forces between them. An important concept here is *binding energy*.

Consider an electron and a proton—the ingredients of the hydrogen atom. Apart and at rest, the system has a total energy:

$$E = m_{\mathrm{p}}c^2 + m_{\mathrm{e}}c^2$$

The mass of the newly formed hydrogen atom m_{H} is *less* than the sum of the masses of its proton and electron ($m_{\mathrm{p}} + m_{\mathrm{e}}$) by a tiny amount, called the energy of binding.

$$
\begin{aligned}
m_e c^2 &= 0.511 \text{ MeV} \\
m_{\mathrm{p}} c^2 &= 938.3 \text{ MeV}
\end{aligned}
$$

$$
\begin{aligned}
\text{Binding energy} &= (m_{\mathrm{H}} - m_{\mathrm{e}} - m_{\mathrm{p}})c^2 \\
&= -13.6 \text{ eV}
\end{aligned}
$$

Why does the mass decrease? Why is the binding energy negative? Note that in order to separate the electron and proton and so destroy the atom, we must clearly put energy into the system, in fact, a minimum of 13.6 eV. This value is somehow related to the size of the atom and the strength of the electric force. In the detailed analysis of the hydrogen atom on the basis of the quantum theory, this number emerges in exact agreement with measurement.

Similarly, when we add neutrons and protons, collectively known as *nucleons,* to a nucleus, the total mass of the nucleus increases, but *not* in the simplest way imaginable. *The total mass of a nucleus is always slightly less than the sum of the masses of the constituents.* This is because some energy was used to bind the constituents into the whole.

In the graph on the previous page, the binding energy B of all nuclei is divided by the number of nucleons A (the atomic weight), and this quotient is graphed against A. We see that after a rapid rise to about 8 MeV per nucleon, the binding energy per nucleon is roughly flat, decreasing slowly as we get to very heavy nuclei. It reaches a maximum at iron ($A \cong 60$). The implications of this curve are tremendous.

If a nucleus of $A = 240$ is split into two products, each of $A = 120$, the binding energy B of the two parts is greater than B of the original nucleus; so the reaction releases energy. Reading from the curve, we see that the binding energy per nucleon of the parent ($A = 240$) is 7.1 MeV, and the binding energy of the daughters is about 8.1 MeV. Recall that binding energy subtracts from rest-mass energy. About 200 MeV is released in the process. This conclusion is the basis of *fission* energy.

Let us survey the sizes of some physical systems and their relevant energies. These representative binding energies, or equivalently, energies required to disturb the systems enough to gain useful information, are shown in the table.

....

BINDING ENERGIES

System	Approximate size (in meters)	Energy
Molecules	10^{-8}	Few tenths of an eV
Atoms (outer electron shells)	10^{-9}	A few eV
Atoms (inner electron shells)	10^{-10}	~1,000 eV = few keV
Nuclei	10^{-14}	Tens of MeV
Neutrons, protons	10^{-15}	~A few GeV
Quark effects	10^{-16}–10^{-18}	100s of GeV and higher

The fission of uranium in an atom bomb produces tremendous energy.

The energy scales refer to the energy per single atomic or sub-atomic system. To emphasize the bridge between microscopic and macroscopic quantities, let's look at a few important reactions. The equation

$$C + O_2 \rightarrow CO_2 + 10\,eV$$

represents the burning of carbon, as is found in fossil fuels such as coal or oil. It asserts that there is a net *surplus* of binding energy when oxygen and carbon combine. This makes carbon a source of chemical energy: 10 eV per reaction of *each atom*.

What do we get from burning a *gram* of coal? There are about 5×10^{22} carbon atoms in a gram of coal; so we release 5×10^{23} eV for every gram of coal that is completely burned! This is slightly less than 10^5 joules, enough to keep a 100-watt lamp lit for about 1 minute at the usual efficiencies.

A nuclear reaction in which uranium nuclei undergo fission re-leases, by contrast, something like 200 MeV per reaction; a gram of uranium that completely fissions would yield about 10^{11} joules, 10^6 times more energy than burning a gram of high-grade fossil fuel. This would keep the lamp on for about 1,000 hours. The energy release in fission follows from the specific structure of nuclei, which is such that the rest-mass energy of heavy nuclei like uranium is *greater* than the sum of the rest-mass energy of the nuclei into which the uranium fissions. The difference comes out as usable energy. In very lightweight nuclei it turns out that energy is *required* to break up a nucleus into parts. But this implies that causing two parts to *fuse* will result in emission of energy. Current research on fusion seeks for a practical way to use this energy. As we shall see, these issues are crucial to understanding how stars work.

FERMI AND THE WEAK FORCE

Attempts to explain certain properties of beta decay led to the recognition by Enrico Fermi of another fundamental force—the weak force. To put this 1933 work into proper context, we must review some of the earlier developments.

In 1927, a young Cambridge theorist, Paul Dirac, attempted the ultimate matchmaker's challenge—to marry Einstein's special theory of relativity with quantum mechanics. He chose as his specific area of investigation the behavior of the electron subject to electromagnetic fields. Prior to Dirac, relativity and quantum mechanics were considered separate revolutions. Quantum mechanics had been applied to slowly moving electrons in atoms. Special relativity showed up when velocities of particles approached the velocity of light. Until Dirac, no one had succeeded in rendering the equations for the electron consistent with relativity. Dirac's efforts led to a new equation governing the behavior of the electron in the presence of fields. The equation, when solved, made several startling predictions: One, already known to be true, was that the electron had "spin," in fact exactly one-half unit of quantized angular momentum (a measure of the rotation activity in an object). The other prediction was totally unexpected: Dirac's equation predicted that a particle of positive charge must exist, with the same mass and spin as the electron. It is typical of a successful theory, such as Dirac's, that the observational consequences of the proposed truth are far richer than the input facts of nature. How can this be? The reason is that the input contains inspired generalizations, in this case that quantum mechanics and special relativity must be compatible. Thus, Dirac predicted the existence of what soon came to be known as *antimatter*.

Dirac's powerful equation demonstrated a complete symmetry between matter and antimatter. This symmetry could be stated (in terms popular in science fiction) in the following form: If all the particles in a system (e.g., laboratory, solar system) were changed to antiparticles, there would be no way for a distant observer to tell the difference. The laws of physics would be the same in the anti-laboratory as in the original laboratory. (Thirty years later, small corrections to this symmetry statement would be discovered.) With Dirac's work on relativistic quantum mechanics in place, physics was now ready for Enrico Fermi.

In the stellar list of inner-space heroes, one star is brighter than most. Enrico Fermi (1901–1954) had the serenity sometimes associated with genius and the grace to surround himself with brilliant students. Fermi was also unique in that his contributions were made equally in theoretical and experimental physics.

In his early work, Fermi gave an interpretation of the Pauli exclusion principle in terms of the statistical nature of electrons. Particles, such as the electron, with half-unit spins are now known as *fermions,* and are said to obey "Fermi statistics," which include the exclusion principle. In contrast, integer-spin particles, such as photons (spin 1) and alpha particles (spin 0), obey a completely different set of rules, called Bose-Einstein statistics. They are known as bosons.

Pauli's 1931 suggestion that the radioactive decay of nuclei is associated with the emission of electrons and neutrinos (i.e., beta decay) made an impression on Fermi, then 30 years old and a professor of considerable international reputation teaching in Rome. Fermi was aware of the work on quantum theory of electromagnetic fields that had been done by Dirac, Pauli, and Heisenberg.

These theorists had shown that the classical force fields are replaced in quantum mechanics by particles that carry the influence of the fields from point to point in space and time. In the case of the electromagnetic field, the force-carrying particle was none other than the photon—Einstein's quantum of light (i.e., electromagnetic) energy. Thus, in the new language of quantum field theory, the photon carries the influence of the field; the force between two charged particles arises because of the exchange of photons. This idea was not only crucial in establishing a correct and very precise theory of quantum electromagnetism (called QED for quantum electrodynamics), but it served as a model for the quantum theory of all the other forces.

Fermi, in 1933, adopted this idea for the beta-decay processes that had been described by Pauli. In trying this he was the first to state clearly that a new fundamental force was acting. It is still called the Fermi interaction by some, but we will use *weak force.*

Fermi's 1933 paper gave a correct description not only of the energy distribution of electrons emerging from decay processes, but also of lifetimes and many other characteristics of radioactive nuclei. When the decay of the muon into an electron and two neutrinos was later observed,

Fermi's theory described it correctly. In fact, the theory was correct for the radioactive decay of all particles. The universal Fermi interaction, as it was called, is the key component in the 1980s standard model. But improved tests of the theoretical predictions of Fermi, Pauli, and Dirac required better instruments.

IMPROVEMENTS IN CLOUD CHAMBERS

If the energetic particle in a cloud chamber happens to collide with the nucleus of the atom, a reaction may take place, and all emerging charged particles will be recorded. Alternatively, a metal plate may be placed across the diameter of the cloud chamber, and the effect on the incident particles observed. One early success of the cloud-chamber technique was the visualization of nuclear reactions, first observed by P. M. S. Blackett in 1932. The reaction Blackett studied was an alpha particle striking a nitrogen nucleus in the cloud-chamber gas, producing an oxygen nucleus and a proton; the reaction is written

$$_7N^{14} + {}_2He^4 \rightarrow {}_8O^{17} + {}_1H^1$$

This alchemist's dream had been surmised from Rutherford's scintillation technique with the zinc sulfide screens, but the cloud-chamber evidence was dramatic.

The power of the cloud chamber as an instrument for observing the microworld grew as a result of two embellishments. One was the idea of placing the entire chamber between the poles of a magnet. In a magnetic field, the paths of charged particles form circles or parts of circles, as in the cloud chamber photo on this page. The radius of the circle or the curvature of the trajectory is a measure of the momentum of the particle: A small curvature means a large radius, with the limiting curvature of zero reflecting an infinite radius, that is, no bending. The higher the momentum, the more difficult it is to curve the trajectory. Hence, a measurement of the radius of curvature on a projection of the cloud-chamber photographic negative can directly give the momentum of the particle.

In 1933, Carl D. Anderson of the California Institute of Technology constructed an extremely strong magnetic field around a cloud chamber in order to produce measurable curvatures in the very high energy cosmic rays that might pass through the chamber. This hot new technique was being attempted in Cambridge, England (naturally), but

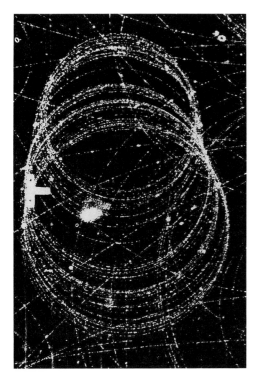

A modern cloud chamber picture of many tracks, including a spiraling electron. A slow-moving particle such as this electron becomes trapped in the magnetic field and travels in circles. As the electron loses energy, the orbit radius decreases.

Carl Anderson is shown working with the electromagnet that will provide a magnetic field for his cloud chamber. The magnet consists of coils of heavy wire wound on a cylinder. There is enough of an opening in the coils that the cloud chamber can be easily photographed.

NUCLEAR NOTATION

$$^A C_Z$$

Z is the number of protons in the nucleus.

A is the number of protons plus neutrons.

C is the chemical symbol.

Example:

$$^{16}O_8$$

The nucleus of oxygen has 8 protons and 8 neutrons.

also in France, Germany, and Russia. However, Anderson at 25, with his own student S. Neddermeyer, was starting a new technique and a new laboratory without the traditions of his experienced European competitors.

The cloud chamber could tell you the sign of the electric charge if you knew the direction of motion. The cloud chamber could also allow you to estimate the *mass* of a particle which leaves a track from the following argument. The higher the velocity of a passing particle, the fewer collisions were made with gas atoms, that is, the fewer the number of ions produced in the wake. Thus, particles moving with high velocities, say, near the velocity of light, have thin tracks with a small number of droplets. Alternatively, slow-moving particles leave behind heavy tracks. Recall that the momentum $p = mv$, and for a given p, if v is very high, as judged by the thinness of the track, then m is low. Since only electrons and protons were known in that idyllic time, it was not too difficult to distinguish them. In Anderson's early photographs, a surprising number of *thin*, low-momentum, positive tracks appeared. (Were they negative, but moving upward? To resolve this issue, Anderson inserted a thin lead plate across the middle of the chamber. This would cause particles which pass through the plate to lose energy or momentum as measured by their curvature before and after the plate.) The photograph that won Anderson a Nobel Prize in 1936, when he was 31 years old, demonstrated the discovery of the positive electron or *positron*. It is

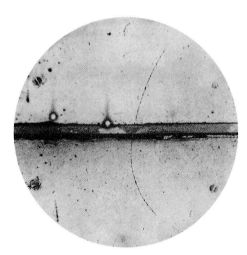

Anderson's prize. This photograph alone established the existence of a positive electron. The direction of motion gives the sign of the charge. Since the particle loses energy as it passes through the plate across the cloud chamber, its curvature is greater afterward. Thus, the particle is moving from top to bottom. That it is a very low mass particle is deduced from the thinness of the ionization along the track.

shown on this page. Anderson's initial interpretation was cautious. Discovering new particles was not an everyday occurrence. (He might have been bolder had he been more keenly aware of Paul Dirac's theoretical work, in Cambridge, in which the existence of the positron or "antielectron" had been predicted.)

The second embellishment to the cloud-chamber technique was made by Blackett, who added Geiger counters to the cloud-chamber arrangements. Recall that cloud chambers are made sensitive by a sudden cooling of the gas, causing a supersaturated condition. Any particle that passes either just after *or just before* the sudden cooling will have its track rendered visible by the growth of liquid droplets on the residual ions. Blackett and a charismatic Italian physicist and speleologist, G. Occhialini, triggered the sudden cooling by electric signals from Geiger counters, indicating that a particle had passed through the chamber. This was enormously more efficient than the previously random photographs taken in the hope that something might happen. Pictures of enormous clarity emerged, quickly verifying Anderson's discovery. The counter-controlled cloud chamber enjoyed an important position in the cosmic-ray and early accelerator studies.

THE CLOUD CHAMBER DETECTS THE MUON

Carl Anderson later adapted the counter-control technique to a new magnetic cloud chamber, which he operated at Pike's Peak in the Colorado Rockies. He was then trying to understand the curious behavior of cloud-chamber tracks at sea level, tracks that did not seem to fit the behavior of electrons or protons. These could not be protons because their ionization tracks were too thin, indicating a higher velocity than a proton of that curvature (i.e., momentum) should have. They could not be electrons because their tracks readily penetrated thick lead plates.

In 1936, Anderson and Neddermeyer finally decided that their data suggested a new particle of intermediate mass between electron and proton. This conclusion was published in 1937, almost simultaneously with similar data from several other groups. Initially the new particle was named "mesotron," indicating its intermediate mass, which turned out to be about 100 MeV. It is now called the *muon*. It is the primary ingredient of sea-level cosmic rays, is capable of penetrating deep underground, but is inert to the strong nuclear force, apparently ignoring it. These conclusions emerged in subsequent cosmic-ray research extending through the 1940s.

The discovery of the penetrating muons in the cosmic radiation by Anderson and others led to an interesting set of errors. In 1935, Hideki Yukawa, a Japanese theorist, had predicted the existence of a particle whose mass would be about 200 times that of the electron. He wanted this "mesotron" to explain not only the strong nuclear force between protons and neutrons *but also* the weak force that induces radioactive decay. Yukawa (1907–1981), making a numerical error, expected the "mesotron" to decay into an electron and a neutrino with a mean life of about a microsecond. Anderson's new particle seemed to fill the bill. The instability of the muons (as they are now called) was suspected from early observations, showing that many more of them disappeared in the time needed to pass through a given mass of air than during the much shorter time needed to pass through an equal mass of solid material like carbon. Why would more of them disappear in the air unless the extra passage of time was significant—that is, unless they were unstable? And if they were unstable, what was their mean life? This puzzle was solved by the work of Bruno Rossi.

ROSSI MEASURES THE LIFETIME OF THE MUON

Bruno Rossi (1905–) was one of the early trickle of refugees from pre-World-War-II Europe, later to grow to a torrent. He fled the anti-Semitism of prewar Italy in 1938, even though he had a considerable reputation as a cosmic-ray expert based in Padua. He found a welcome first in Blackett's laboratory in England and then in the United States at the University of Chicago. Later at Cornell and at M.I.T. he led teams that greatly aided the unraveling of cosmic-ray phenomena. Rossi was also an expert in the new subject of electronics, the art and science of combining data from electrical detectors such as Geiger counters to define trajectories in very precise time frames. Rossi was one of the pioneers in inventing circuitry that later became the mainstay for accelerator research. These early circuits, invented for cosmic-ray and nuclear research, became the essential elements for the electronic revolution including digital computers, which now dominate scientific instrumentation and life in general.

Rossi's coincidence circuit is shown on this page. If a fast particle passes through two Geiger counters sequentially, the electrical impulses arrive at the circuit essentially simultaneously. The circuit has an output *only* when there is such a transit and, in turn, the output pulse indicates the passage of a particle.

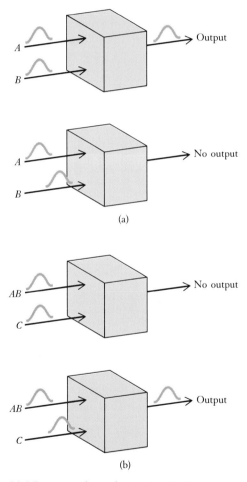

(a) The coincidence logic circuit: An output pulse occurs only when electrical pulses *A* and *B* are simultaneous. (b) The anticoincidence unit: An output pulse occurs only if there is no *C* pulse simultaneous with the *AB* signal.

The Geiger-counter setup for detecting cosmic rays. In this arrangement, a particle coming down from above and stopping in block II would give rise to electric pulses in Geiger-counter trays *A* and *B* but not in tray *C*. The circuit would give an output for only such events, which signal that a charged cosmic ray has stopped in block II. Cosmic ray muons pass easily through the lead of block I, which filters out other particles.

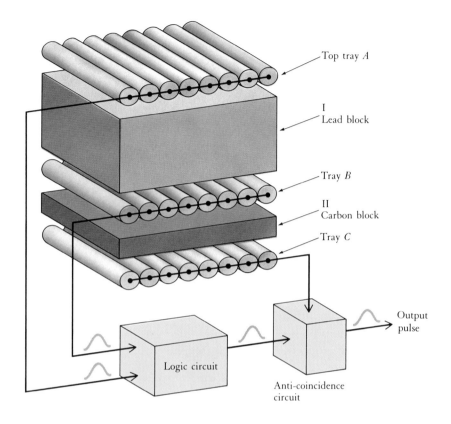

Top tray *A*

I
Lead block

Tray *B*

II
Carbon block

Tray *C*

Output pulse

Logic circuit

Anti-coincidence circuit

By increasing the number of counters required, we can even define the direction of incoming particles. Later Rossi invented an anti-coincidence circuit (shown on this page), which informs the physicist that, for example, a particle passed through trays of Geiger counters labeled *A* and *B* but *did not* pass through tray *C*. This type of circuit is often used when a carbon block has been placed between trays *B* and *C*. If the circuit sends a signal, then the particle must have come to rest in the carbon block. This latter circuitry is a good example of the increasingly sophisticated information that arrangements of counters and electronic read-outs can supply. As a prototype of this, and one of the most elegant of the cosmic-ray experiments, we will consider the direct observation of the lifetime of the muon, measured by Rossi in 1943. But first we must say something about the nature of radioactivity and the instability of particles.

We have already noted that the spontaneous disintegration of particles is brought about by the action of the weak force. This influences all the elementary particles except those of the very lowest mass. The lifetime of an unstable particle (or nucleus) is defined as the time necessary for a certain fraction of the particles to disintegrate. Given a certain number of particles, for example, muons, the *half-life* is the time it takes for half of them to decay. A more convenient measure is the *mean-life* or the time it takes to leave only 37% undecayed.

Suppose we have only one muon. What can we say about when it will disintegrate? It turns out that the same formula applies, except that we now interpret the 37% as the *probability* that the muon is still around after one mean-lifetime has passed.

If we can measure the time at which each of a large number of muons decay, we can calculate a mean-life by a suitable averaging. Rossi knew that one of the decay products of a positive muon is a positive electron, or positron:

$$\mu^+ \rightarrow e^+ + \text{neutral things}$$

A Geiger arrangement similar to Rossi's 1942 experiment on page 68 was used to record those events in which a cosmic-ray muon would come to rest in a *carbon block*. This used both coincidence and anticoincidence circuits. Then Rossi invented a new "timing" circuit, which, in simplified form, measured the time interval between the coincidences of counters in trays A and B and the counts recorded, *shortly after*, in the counters of tray C. These were assumed to be the positron-decay product of the arrested muon. Here is the interpretation of the logical ordering of pulses: A and B fire; nothing fires tray C simultaneous with A and B; and, in a defined interval of time subsequent to this (e.g., 0.1–10 microseconds), there is a count in C. All of this is read to mean that a cosmic-ray muon entered the apparatus, traversed A and B, stopped in the carbon block, and subsequently decayed, giving rise to a positive electron, which fired counter C. The delay between signal AB and C is measured. The result is shown in the graph on the facing page and yields a lifetime for the muon of

$$T_\mu = 2.30 \pm .17 \text{ microseconds}$$

Subsequent research in the late 1940s on the details of the muon decay led to the conclusion that the true reaction was:

$$\mu^+ \rightarrow e^+ + 2 \text{ neutrinos}$$

Of course the symmetry of matter and antimatter implies that

$$\mu^- \rightarrow e^- + 2 \text{ neutrinos}$$

and this was soon verified.

Rossi found that two-thirds of all muons decayed within 2.30 microseconds. The number of anti-coincidences (the vertical scale) gives the number of muon decays recorded. The delay between signals (horizontal scale) gives the time it took for a decay to occur. Two-thirds of the anti-coincidences showed a delay of 2.30 microseconds or less (color shading).

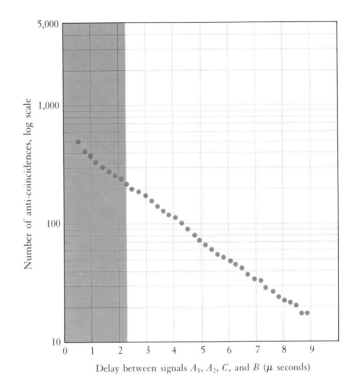

NUCLEAR EMULSIONS AND THE DISCOVERY OF THE PION

Recall that in 1935, in order to account for the strong force, Yukawa had postulated the existence of a particle with about 200 times the mass of an electron. He came to this conclusion by considering the strong-force field between the nuclear inhabitants: neutrons and protons. Guided by the work of Dirac, Pauli, and Heisenberg, who had made a quantum theory of the electromagnetic force, Yukawa noted that the quantum of the strong force that bound neutrons and protons could be a new particle. Its mass of about 100 MeV was related to the range of the force.

When Anderson announced the discovery of the muon in 1937, most physicists believed it to be the Yukawa particle. However, the first discrepancy appeared in 1939. Yukawa's theory, after correcting a mathematical error, predicted that a proper strong-force particle should have a mean-life of about 10^{-8} seconds. As we have seen, by 1944 the muon's mean-life had been determined to be 2×10^{-6} seconds. A second discrepancy also appeared in the early 1940s. The Yukawa particle, being a creature of the strong force, should have a large probability for making catastrophic collisions with nuclei, that is, collisions in which the particle would convert its mass into energy. Thus, Yukawa particles should be easily absorbed. But muons were observed to pass through very thick absorbers without engaging in the strong-force collisions.

The resolution of this crisis was the discovery in 1947 of the pion, which had all the right properties, and the discovery that muons are the decay products of pions. This later discovery came about in the following way.

Radioactivity was found by the observation of the blackening of photographic plates. In the 1940s this accidental discovery was understood and exploited to obtain photographs of incredible clarity. The passage of charged particles through material leaves a trail of ionization. In photographic emulsion, silver bromide molecules are dispersed in a gelatin. In its normal use in cameras, light striking the silver bromide produces a chemical change. When the emulsion is treated chemically, the exposed molecules are reduced to silver atoms, which are opaque to the passage of light. Actually, the silver halides are distributed in small clumps or grains, and the incident light, followed by chemical development, renders the entire grain opaque. The same process takes place in the passage of a charged particle: Its path, after development, is recorded by a trail of grains or clusters of atoms, which are about 1 micrometer in size, that is, about one-thousandth the size of the 1-millimeter drops in the cloud chamber. To see and measure these tracks one needs a strong microscope with a precision stage. The tracks of particles, their collisions, and their spontaneous disintegrations can then be recorded with exquisite precision. The 1950s saw many laboratories occupied by roomfuls of young women scanners, hunched over microscopes, painstakingly measuring subnuclear events.

Perhaps the greatest success of nuclear emulsions was the discovery of the pi meson or pion. Cecil Powell (1903–1969), a graduate of Rutherford's incredible school, set up shop in Bristol where he was joined by G. Occhialini in 1945. Occhialini had been discussing improving nuclear emulsions with the English manufacturer Ilford Ltd. The problem was to increase the number of grains developed along the track of a fast-moving (relativistic) particle. In the then-standard emulsions, too few grains were developed to allow a relativistic track to be distinguished from the normal background of randomly distributed grains. Driven by the pleas of Occhialini, Ilford finally improved the quality of these photo emulsions so that high-velocity particle tracks were clearly visible. Occhialini brought a small number of plates, each measuring 2 centimeters by 1 centimeter with emulsions of a 50-micrometer thickness, to the cosmic-ray station of the Pic du Midi in the Pyrenees mountains on the French–Spanish border. At an altitude of 3,000 meters, the cosmic-ray

The Pic du Midi station in the 1940s, at the time Occhialini discovered the pion.

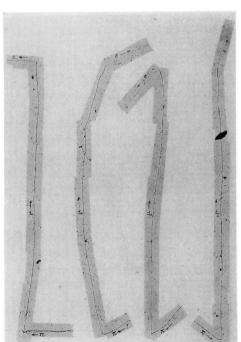

Nuclear emulsion photomicrographs of decaying pions. The incoming pions (bottom) come to rest, and within 10^{-8} seconds decay into muons and neutrinos. The muons (going up) travel a length of 600 micrometers, whereupon they come to rest and within 2 microseconds decay into electrons and two neutrinos.

intensity at the station was ten times higher than at sea level. The new, sensitive emulsions were exposed on the mountain top, developed, and brought to Bristol late in 1946, where many measuring microscopes were available. The scanners who operated these microscopes were trained to locate and measure interesting tracks. Occhialini supervised the scanning. (It is told that an Italian friend, just arrived from Rome, was searching for Occhialini when he wandered into a room filled with young English women, innocently using the most profane Italian expletives. The visitor knew he had found Occhialini!)

The plates were a bonanza of never-before-seen activity. The power of the sensitive emulsion opened a new domain, which, magnified through the microscope, showed thin tracks of very high energy particles and thick, black tracks of slow-moving particles. It was in this frenzy of searching that the event illustrated in the figure was found. The interpretation, quickly confirmed by other events, was that a particle of intermediate mass, about 140 MeV, is gradually slowed down by its many collisions with the atoms in the emulsion, comes to rest, and then spontaneously disintegrates into a different charged particle. Later analysis identified the second charged particle as the muon. The first particle, seen spontaneously disintegrating into a muon, was named pi meson and later *pion*. The pion was soon identified as having the properties predicted by Yukawa's 1935 theory. The reaction that led to the discovery of the pion (π) is:

$$\pi \rightarrow \mu + \text{neutrino}$$

The mean-life of the pion was later found to be just that predicted by Yukawa. The strong emission and absorption of the pions is said to mediate the force that holds the protons and neutrons together in the nucleus.

The disadvantage of using emulsions is that there is no time information; the tracks in emulsion begin accumulating the ever-present cosmic radiation from the moment the emulsions are poured. Also, although emulsions as thick as 1 millimeter could be deposited on glass, this did not represent a large volume. To compensate, large stacks of emulsions were exposed, at first to cosmic radiation, and later in accelerator beams. Here again, many important discoveries were made, and although the peak of nuclear-emulsion activity was in the period between 1940 and 1960, the discovery of very short lived particles brought the technique back in the 1980s.

THE BUBBLE CHAMBER

In the 1950s, accelerators began to dominate the field of particle physics, and detectors designed specifically for this operation faced new issues. They had to be able to handle higher-intensity beams, and they had to be ready to track a new beam cycle once every couple of seconds. In 1956, a new visual detector called a *bubble chamber* was invented that tracked particles traveling through a liquid medium.

Using a higher-density liquid medium increases the probability of seeing interesting collisions. The principle is the same as in the cloud chamber, except that bubble formation replaces droplet formation. A pot of liquid can be superheated, that is, heated beyond its normal boiling point, but if it is very clean, it refuses to form bubbles. Again, ionization (instead of dust) provides the required nucleation center around which bubbles can form.

The invention of the bubble chamber was not easy. Its inventor, Donald Glaser, spent years trying different liquids and different techniques for superheating. (He is said to have been inspired by the important act of pouring salt in beer to improve the head in bars outside the University of Michigan campus.) The bubble chamber dominated particle instrumentation from about 1960 to the mid-1980s. Early bubble chambers used liquid propane and pentane as the medium, and had diameters of 5 to 10 inches. These were also equipped with magnetic fields to enable the measurement of the momenta of particles. Later, huge chambers, as large as 5 meters in length and using liquified hydrogen, were constructed. Typically, a bubble chamber would be placed in a beam of particles emerging from a particle accelerator. The number of

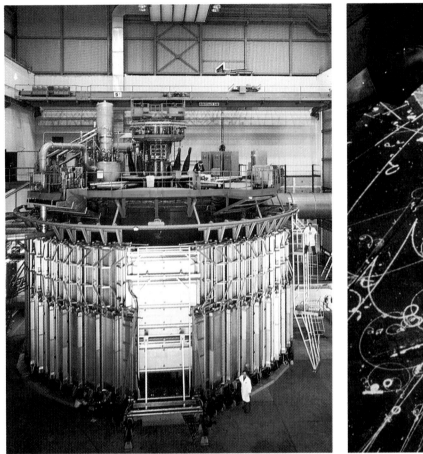

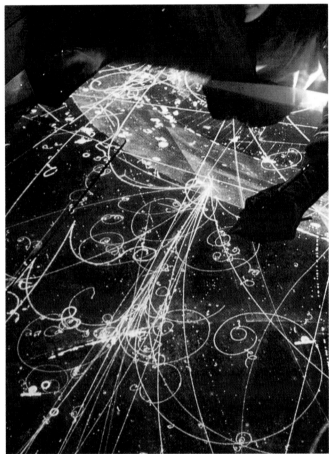

Left: The 3.7-meter (12-foot) CERN bubble chamber, dismantled in August of 1984. One of the biggest in the world, it took over six million photographs showing collisions of neutrinos with the liquid hydrogen nuclei. *Right:* A scanner measures neutrino-induced collisions in Fermilab's 15-foot hydrogen-filled bubble chamber. The scanner places a cursor on many points along a track and, by pushing a button, provides digital information on the location of each point. Computers then give back information on the energy, momentum, mass, etc., of the collision products.

particles in the beam would be adjusted so that the number of collision tracks in each photograph was as large as possible without cluttering the view. For incident charged particles, this number was typically between ten and fifty. Because neutrinos leave no tracks, some millions of neutrinos could be aimed at the target per cycle. About once per minute flash tubes would brightly illuminate the chamber, and sophisticated cameras would photograph the tracks. Some chambers recorded the tracks more frequently. At the peak of bubble-chamber use, millions of such photographs were taken each year, and each was carefully examined to determine the properties of incident particles and the details of the collisions. In the 1960s automation was added to the measuring technique. Photographs could be quickly scanned by a flying spot that registered individual bubbles and passed this information on to computers that then

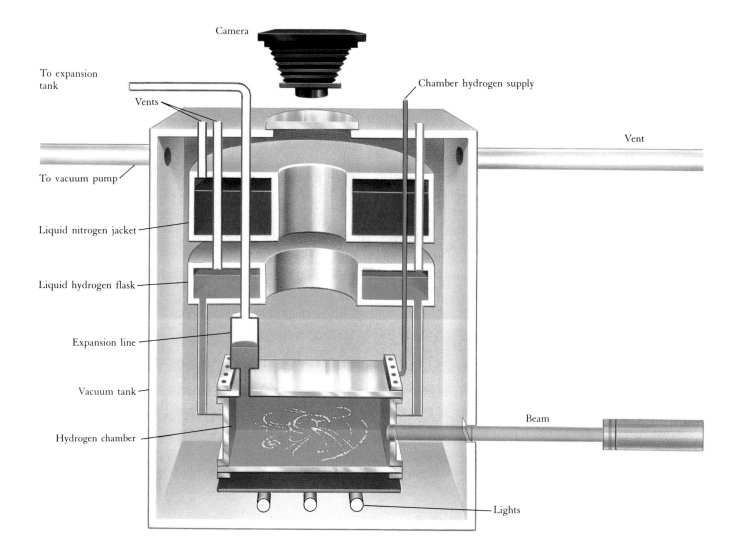

Camera

To expansion tank

Chamber hydrogen supply

Vents

Vent

To vacuum pump

Liquid nitrogen jacket

Liquid hydrogen flask

Expansion line

Vacuum tank

Beam

Hydrogen chamber

Lights

A 10-inch-diameter hydrogen bubble chamber. Just before the particle beam arrives, the expansion valve is operated to decrease pressure on the liquid hydrogen. Instead of boiling, bubbles form initially only along the passage of a particle. When the bubbles are about 1 millimeter in diameter, the camera takes a picture. To keep the hydrogen cold and liquified, the entire system is surrounded by a vacuum tank and a liquid nitrogen shield. After the photograph is shot, the liquid is recompressed, dissolving all bubbles, and the film moves, leaving the chamber ready for the next beam.

extracted only the useful information. (However, even the best of these devices required human guidance to keep the computerized scanning honest.)

Both the Wilson cloud chamber and the bubble chamber had the virtue that the track was accurately delineated, the bubbles being about 1 millimeter or less in diameter. This was important not only for studying the details of the collision process, but also because the ability to accurately measure momenta by the curvature of tracks in the magnetic field greatly enhanced the power of the bubble chamber. In the late 1980s, the last large bubble chamber was honorably retired. By this time holographic photography had been used to improve the sharpness of the bubbles and, hence, the precision of spatial measurements.

SCINTILLATION COUNTERS
AND MWPCS

While we are considering particle-detection techniques, let us cover current developments as well.

The modern version of Rutherford's scintillation screen is a fluorescent chemical embedded in a clear plastic, such as Lucite or polystyrene. Again, the passage of a charged particle creates a tiny scintillation. This very small amount of light is conducted toward the end of the plastic, which can easily be machined to any suitable size and shape. At the end of the plastic, the scintillation counter is cemented to a photomultiplier tube, which greatly amplifies the very tiny amount of light released by the charged particle. The response is an electric pulse signal-

Left: A physicist installs multiwire cables in a Fermilab experiment. Behind is an array of scintillation counters with photomultiplier tubes on top. *Right:* A scintillation counter. Passage of a charged particle through the scintillator releases a series of photons along the path. The front surface of the photomultiplier is a photo cathode coated on the inside with material that readily emits electrons when bombarded with photons. The electrons are focused by electric fields onto a series of "dynodes"; each dynode amplifies the signal so that after 20 dynodes there are over 10^6 electrons. A fast electrical pulse results.

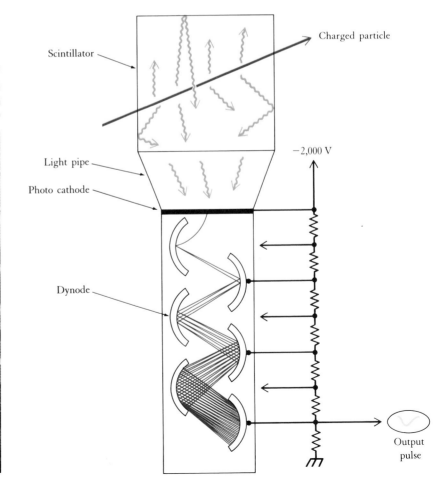

ing the passage of a particle. The total duration of the electric pulse can be as short as a few nanoseconds (10^{-9} seconds). Electronic processing of the information can delineate the time of the particle's passage to a precision of fractions of a nanosecond. Arrays of scintillation counters are now standard items in accelerator experiments, principally because of the very good *time* definition of the scintillation pulse. In accelerator work the intensity of radiation is so high that precise time definition becomes essential.

A dramatic breakthrough in the quest for both spatial and temporal definition of particles was the multiwire proportional chamber (MWPC), invented by Georges Charpak in 1969. In the MWPC many parallel wires are stretched across an insulating frame, every other wire carrying a positive voltage of a few thousand volts. The wires in between are at ground potential. Wires can be as close as 2 millimeters. The plane of wires is enclosed, front and back, in conducting material (sometimes insect screening) and the entire assembly is made gas tight. Passage of particles is registered by small pulses on the charged wires, much as in the Geiger counter. Recall that in the Geiger counter, the initiation of any charge creates an avalanche so that the output loses all memory of the input. In the proportional mode no avalanche takes place, giving the chamber a much more rapid response time and an output that carries information about the strength of the input. A summary of some of the useful detectors is given in the table.

Left: The UA1 collider detector at CERN. In the center of this device took place proton–antiproton collisions at a total energy of 600 GeV. *Right:* The multiwire proportional chamber. In this example, the wires in the central plane have alternating positive and ground potential relative to the top and bottom planes. Electrons released by a passing particle drift to the nearest central positive wire, multiply near the wire, and produce a negative pulse on the wire. The negative pulse indicates one coordinate of the track, the coordinate perpendicular to the wires. This signal induces pulses of opposite sign in the adjacent wires and on the top and bottom wires; the strongest pulses in these planes give the other coordinate of the track segment. The precision is given by the wire spacing, typically a few millimeters.

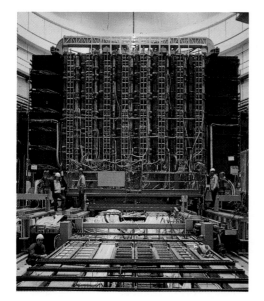

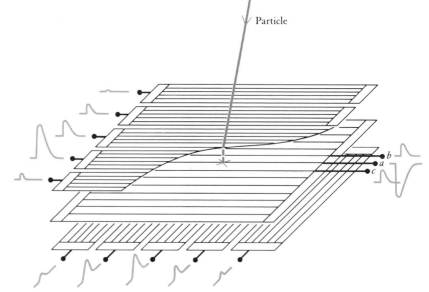

USEFUL DETECTORS

Detector	Time resolution* (nanoseconds)	Space resolution† (micrometers)	General comments
Bubble chamber	None	~500	Detailed view of collisions.
Scintillation counters	~1	Generally ≥ few 1,000	Easily shaped by machining plastic material. Usually used in arrays that can cover many meters.
Multiwire proportional chambers (MWPCs)	~10	100 1,000	Large arrays possible. Wire spacings of ~ few mm. Wire lengths of ~1–2 m.
Cerenkov counters	~1	Typically poor	Used for particle identification. The lower the mass, the more Cerenkov light is generated.
Calorimeters	~100	≥ 1,000	Composite layers of absorber and scintillator to measure energy.
Silicon microvertex detector (SMD)	~Few	< ~10	Generally covers limited area, e.g., 10 cm × 10 cm but with strips that are 20–50 μm wide.

*This is the time definition used in relating one event to another. If, for example, two "hits" in an MWPC are separated by more than the time resolution, it is likely that they came from independent events and should not be correlated.

†A smaller value enables one to make precise momentum and angle measurements, or to resolve events very close to the vertex of a collision.

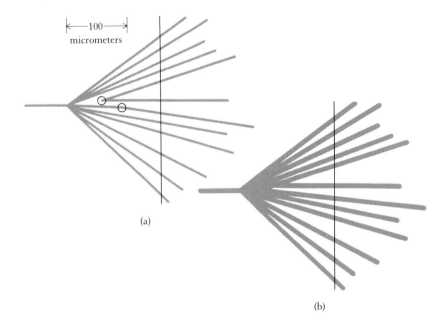

(a) A high-energy collision produces two unstable particles, seen decaying at the spots marked by the circles. The tracks indicating the decays are clearly recorded by detectors with high spatial resolution. (b) The same event recorded with poor spacial resolution. It is obvious that a detector with poor spatial resolution would miss these short-lived events.

NUCLEAR ASTROPHYSICS

Progress in detectors was both motivated by and applied to problems in astronomy. By the end of the 1930s, the increasing understanding of nuclear physics was employed to solve a long-standing problem of astronomy; namely, What provided the energy for stars?

The basic structure of stars had been deduced by the beginning of the twentieth century by such astronomical theorists as Sir James Jeans and Lord Kelvin. These scientists realized that a mass of gas pulled together by its own gravity would have its center heat up as it became very dense. This is just the conversion of gravitational potential energy to kinetic energy of gas molecules as the gas "falls" toward the center in the process of forming a star. The high energy in the core would gradually radiate out through the gas and make the surface of the gas ball very bright. However, the same studies also revealed that if gravitational contraction were the only source of energy, the star would continue to collapse. It was estimated that all of the energy in a star like our Sun would radiate out in about 10 million years. When Lord Kelvin first pointed this out, geologists were gleeful in noting that the solar system must be much older than 10 million years, since there were rocks on Earth far older. Obviously there needed to be an additional energy source for stars beyond that generated by their gravitational collapse. Such an energy source would enable stars like our Sun to continue to burn past the 10-million-year gravitational timescale. The longevity of the Sun was thus a major problem.

The solution was that stellar energy came from nuclear reactions. We have seen that the binding energy per neutron or proton decreased for elements lighter than iron. This meant that when two light nuclei combined to make a heavier nucleus (which was still lighter than iron), the heavier nucleus was more tightly bound and the reaction released energy. Such a reaction is called *fusion*, since two light things are fusing together to make a heavier one. On the other hand, nuclei heavier than iron would want to spontaneously break up into lighter nuclei, getting closer to iron, where the maximum binding energy per nuclear particle exists. This breakup of the heavy nuclei to lighter ones is called *fission*.

Stars are primarily made up of hydrogen, with some helium, and only traces of heavier elements. Thus, as the core of stars gets hot, the hydrogen wants to fuse together to make heavier things. The details of this mechanism were first worked out by Hans Bethe in 1938. In particular, Bethe showed that the fusion of four hydrogen nuclei to make a helium-4 nucleus could explain the energy source for the bulk of the stars in the sky, the so-called main-sequence stars.

It is in interstellar gas clouds such as this one that new stars are formed from gas contaminated with the debris of old stars rich in heavy elements.

Subsequently it was shown that another, rarer class of stars, known as red giants, could be explained by the fusion of three helium nuclei to form a carbon-12 nucleus. In one of the first examples where an astronomical argument was used to tell us something about laboratory physics, Fred Hoyle argued in 1954 that three helium nuclei could combine to create carbon 12 only if there was a particular energy level of carbon 12 at a little over 7 MeV above the carbon-12 ground state. At the time of Hoyle's prediction, no such level was known. It was subsequently searched for in experiments done by William A. Fowler and his group at the California Institute of Technology in 1957. To their astonishment, the energy level of carbon 12 was exactly where Hoyle had predicted. Thus, helium really could combine to make carbon and provide the energy source for the cores of red-giant stars. Fowler (1911–) and Hoyle (1918–) continued to follow nuclear-reaction networks in stars, and were able to explain the synthesis of elements by subsequent burning stages all the way up to iron. By the late 1950s Fowler, Hoyle, and Al Cameron (now at Harvard) were also able to show that if neutrons are added to iron, elements beyond iron are made.

Neutrons are needed to create elements beyond iron because iron has the maximum binding energy per nuclear particle, and fusion reactions cease at iron. Charged-particle reactions would not work effectively to get beyond iron because the proton, or any other charged particle, would be repelled by the high positive charge of the iron nucleus. Only very high energy protons could overcome the electric-charge barrier put up by the nucleus. Very few such energetic protons existed at the red-giant stage of the star's evolution. However, if neutrons were present, they might sneak in, not being repelled by the positive charge of the iron nucleus, collide with the iron nucleus, and stick, thus building up elements heavier than iron. These heavier nuclei would initially be unstable, and would decay via beta-decay processes into stable nuclei with more protons than iron. Red-giant stars had reactions that could produce some neutrons that would be captured by any iron that existed on those stars, and could thus produce some of the elements beyond iron. Some red giants did indeed show enrichment in these elements.

An alternative source for the neutrons would be the explosions of supernovas. In these explosions, neutrons might also be produced and could make some of the elements beyond iron. In addition, the supernova explosions would throw out all the heavy elements that were synthesized in the various burning stages: helium, carbon, oxygen, neon, silicon, and iron. The supernova that blew up in the Large Magellanic Cloud and was first observed on February 23, 1987, probably threw out the equivalent of about five solar masses of heavy material between carbon and iron. Five solar masses of heavy elements is enough to make 5 million Earths.

SUMMARY

New devices for detecting particles led to a vast number of discoveries in the period between 1930 and 1940. The properties of the nucleus and the power of induced nuclear reactions greatly clarified the domain of the very small—the realm of 10^{-15} meters. The use of this domain to explain how a star works was, along with the cosmic rays, another example of the connection between inner and outer space. Out of this period came a clear awareness that the inner space of the nuclear world contained two new, fundamental forces, the strong and weak forces, that would join the electromagnetic and gravitational forces as the basic laws of nature.

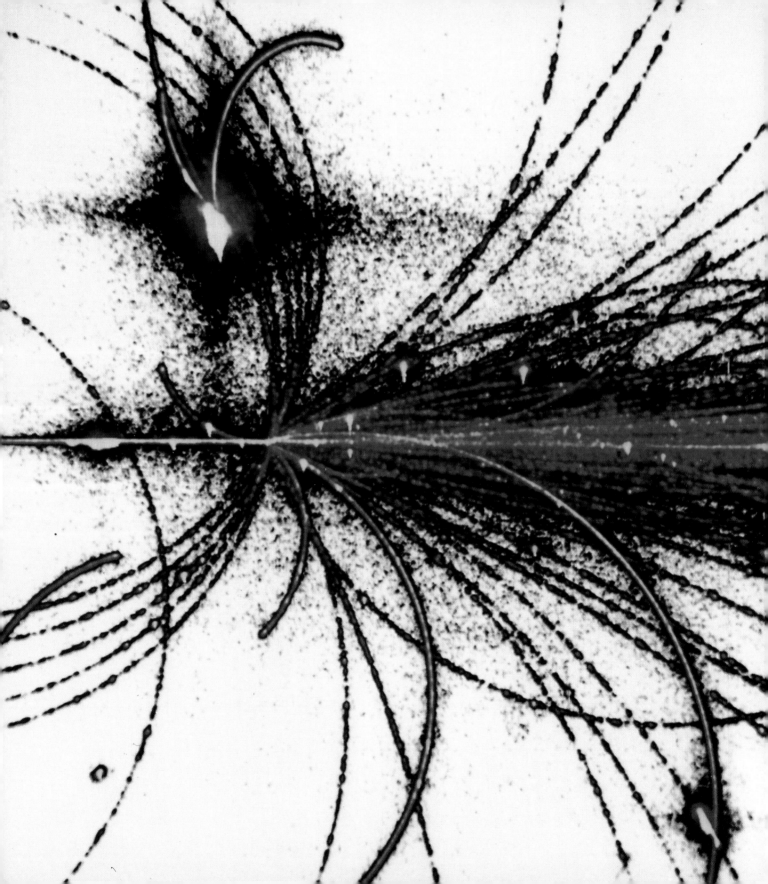

4 PARTICLE ACCELERATORS AND THE STANDARD MODEL

A computer playback of a very high energy oxygen nucleus striking a neon nucleus. The collision occurred in a streamer chamber, which contains a very high voltage that produces tiny sparks along the ionization trail of fast-moving particles. A magnetic field allows the momentum to be measured. The camera replaces photographic film with electronic charge-coupled devices (CCD), which can feed data directly to the computer.

We now know that in the maelstrom of high density and temperature that existed in the early moments of the universe, only the most primordial objects could exist; any transient combinations, such as protons, let alone molecules, would decompose more quickly than a butterfly in the core of a volcano. It is for this reason that the search for what is elementary underlies both particle physics and the cosmology of the early universe. It is also true that what is considered elementary changes with the quality of our observing instruments.

Dmitri Mendeleyev's periodic table of the elements served as an ordered list of all the known "atoms," and its mysterious regularities could be used to predict the existence of as-yet-undiscovered elements. It was not until the discovery of the electron and of quantum mechanics that the "why" of the periodic table became understood. In the 1980s, the corresponding organization of basic particles is called the standard model. It contains three classes of particles: quarks, leptons, and gauge bosons.

Both quarks and leptons seem to be elementary and indivisible. *Quarks* are the constituents of protons and neutrons and of *mesons*—the Yukawa particles that keep the protons and neutrons together. Quarks are bound together to make these particles by means of the strong force. In contrast, the *leptons,* which include electrons, muons, and neutrinos, are not affected by the strong force. The forces between elementary particles are carried by the *gauge bosons.* In this chapter we will survey the experiments that have led to the construction of this model. The key tool in all of these experiments was the particle accelerator.

TYPES OF PARTICLE ACCELERATORS

In 1919, Rutherford demonstrated that the nitrogen nucleus could be disintegrated when bombarded by alpha particles. This opened up a new field, nuclear physics, and a new level of observation. The race to higher energies began in "Ernest."

The ability of very high speed particles to produce nuclear disintegrations led to attempts to produce particles that are even more energetic than those that emerge naturally from radioactive substances. It was obvious that if a device could accelerate particles and even generate an intense stream of very energetic particles, it would be enormously useful for studying the properties of nuclei.

The simplest way to accelerate a particle is to use the potential of an electric field to increase the velocity of a particle that has an electric charge; and, in fact, particle accelerators can accelerate only charged particles. The acceleration occurs when the moving particle crosses a *gap* in the accelerator, where it is pulled ahead by a charge of opposite sign, and usually pushed from behind by a charge of the same sign (see the figure on this page). Every time the particle crosses a gap, it is accelerated by the electric kick it receives.

The primary element in particle accelerators is the gap; when crossing the gap particles gain energy equivalent to the voltage applied. Some source of particles is required, here a hot filament that emits electrons. Some way of selecting a narrow beam is needed, here a slit. Electrons emerging from the slit are attracted across the gap by a screen at positive high potential. Most of the accelerated particles pass through the screen.

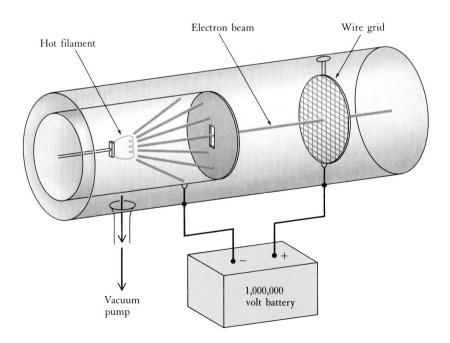

Hot filament   Electron beam   Wire grid

Vacuum pump

−   +

1,000,000 volt battery

A beam of particles is scattered by a target. The electronic counter will detect all scattered particles heading at an angle θ with respect to the original beam. Detectors at accelerators frequently contain thousands of counters at different angles.

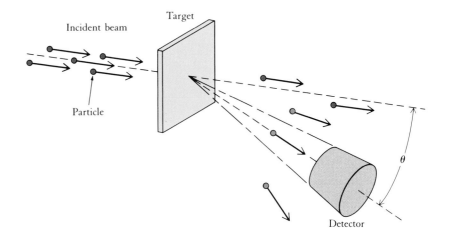

Beginning from the simplest applications of this principle, particle accelerators evolved from the relatively small ones of the 1920s to the giant machines of the 1980s, which have achieved energies of almost a million times that of Rutherford's alpha particles.

What happens when the particle has reached the maximum energy to which the machine can accelerate it? That depends on the design of the machine. There are two basic designs, *fixed-target* machines and *colliders*. In a fixed-target machine, the accelerated particles are extracted from the accelerator and guided to a place where they strike a fixed target, usually a cylinder of metal. Some of the particles will strike the nuclei of the metal atoms and will, if they are energetic enough, knock loose a spray of neutrons, protons, and other particles, all of which can be tracked by suitably placed particle detectors and analyzed by computers (see the figure on this page). Sometimes the spray of particles can be formed into a secondary beam, which strikes another target, and so on.

In a prototypical arrangement, a stream of (for example) protons from a particle accelerator is aimed at a thin carbon plate located across the diameter of a cloud chamber. Most of the protons miss the tiny carbon nucleus, and pass straight through the plate. Some graze the nucleus and are deviated slightly. Still more rarely, a proton strikes the nucleus directly.

Various results can follow, depending on the energy of the proton and on the details of the strong force between the proton and the particles in the carbon nucleus. By compiling the results of thousands of such collisions, we have gathered data on the nature of the forces and the properties of the particles. The energy of the bombarding particle is the most important parameter in analyzing these results.

In colliders, two separate batches of particles revolve around the accelerator in opposite directions. Since each batch is accelerated to the maximum velocity the machine can achieve, their head-on collisions occur at twice the energy achievable in a fixed-target mode. However, the gain in collisional energy in the head-on mode is far greater than this mere factor of two. This comes from the law of conservation of momentum. In the fixed-target example, the particles flying off after the collision must have kinetic energy in order that their total (vector) momentum adds up to the momentum of the incident particle. This kinetic energy subtracts from the collisional energy. Suppose, for example, the goal is to create a new heavy particle X. In the fixed-target mode we need a lot *more* energy than $m_X c^2$ (the product of X's rest mass and the speed of light squared) since the X particle must also have kinetic energy. In a collider, the total momentum is zero since the momenta of the objects colliding head-on are equal and opposite, and therefore cancel. Thus, *all* of the energy of *both* colliding particles may be used, for example, to create new particles. As a numerical example, the 1,000 GeV protons at Fermilab, working in the fixed-target mode, have 42 GeV of collisional energy available, but in the collider mode, there is 2,000 GeV available.

Each head-on collision of two particles creates a spray of secondary particles, which again can be tracked and analyzed. The most common combinations of colliding particles are electrons and positrons, protons and antiprotons, electrons and protons, and protons and protons.

Each type of accelerator and choice of colliding particles has its own problems and advantages. Which type is used depends on the specific research problem under investigation, as we will see. Our knowledge of the nature of matter depends not only on the high energies to which the accelerators can raise particles, but also on the sophistication of the particle detectors that have been developed in recent decades, because it is the detectors that tell us what has happened when accelerated particles collide.

THE EARLIEST ACCELERATORS

Rutherford's skillful use of naturally produced alpha particles stimulated a race for more convenient probes, particles that would emerge from accelerators at high speeds. The first accelerators, as we noted in Chapter 2, were the cathode-ray tubes, which accelerated electrons to several thousand volts in evacuated glass tubes. Heinrich Hertz and Philip Lenard in 1903 stepped up the energy to several hundred thousand volts, and saw the electrons penetrate metal foils. Lenard did the first scattering experiments with artificially accelerated particles.

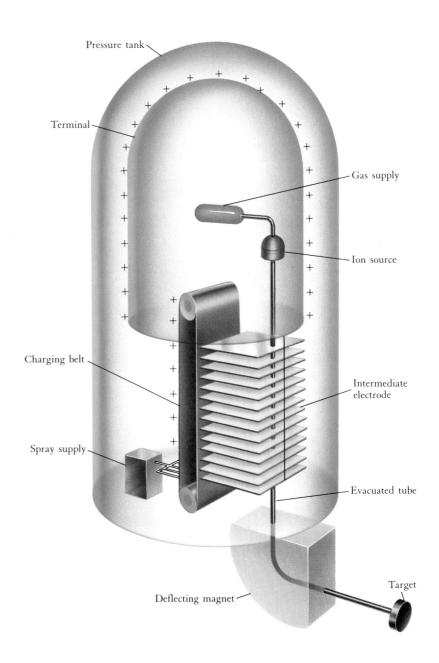

A Van de Graaff generator produces accelerated particles in the form of positive ions. To make the ions, a spark in the ion source strips electrons from the nucleus of a gas such as hydrogen. To discourage further sparking, the ions are born inside a high-pressure tank. A mechanical belt carrying electrical charge gives the tank a voltage of anywhere from 0.1 to 10 MeV. The ions are accelerated down the evacuated tube, gaining the energy at the terminal. A deflecting magnet aims the ions at the target.

The engineering problem of achieving voltages comparable to the 5-MeV alpha particle was formidable. The first success took place at the Cavendish Laboratory in 1932, where John D. Cockcroft and Ernest T. S. Walton succeeded in accelerating protons to 770 keV in an electrostatic machine, a machine making use of a fixed, steady voltage. These protons were used to observe the disintegration of lithium nuclei. A theorist of

The Cockcroft–Walton electrostatic pre-accelerator at CERN. The column in the foreground generates 800 keV. It is connected to the housing (square box), which contains a source of hydrogen gas and a spark generator to produce a beam of protons. The protons are accelerated from the 800-keV terminal to ground potential and enter the linear accelerator.

that day, George Gamow (whom we will hear more about later), had calculated that it would take 770 keV of energy to overcome the coulombic repulsion between the positive lithium nucleus and the incoming positive proton.

The most successful electrostatic machine was developed by Robert Van de Graaff in 1931. He invented an "endless" insulating belt that carried electric charge up to a large insulated metal sphere, as shown in the diagram on the previous page. The machine was cheap and easy to build, and reached 1.5 MeV. This was widely copied and continuously improved upon as the need for higher energies for nuclear bombardment became urgent.

E. O. LAWRENCE AND THE CYCLOTRON

The technological breakthrough that eventually gave birth to modern particle accelerators was made by Ernest O. Lawrence (1901–1958) in Berkeley, California, around 1930. He received the Nobel Prize in 1957 for the application of this entirely new principle of acceleration, and in his Nobel lecture he detailed his discovery and its inspiration. In 1929, as a Berkeley professor, he was interested in studying nuclear reactions, and considered various forms of high-voltage generators. He accidentally ran across an article by a Norwegian engineer, Rolf Wideroe, who had considerable experience with various forms of particle accelerators. Wideroe, a self-taught genius, had at the age of 21 invented a magnetic-induction accelerator, that is, a device that substitutes a time-varying magnetic field for the electric-field "gap" described earlier. In 1925, Wideroe had designed one that would accelerate to 100 MeV. He never got to build it, but judging from his plans we can suppose it would have worked twenty-five years ahead of its time. The key idea here was to use more than one gap, or use the same gap over and over again. Wideroe had written up some of his ideas on multiple accelerations, and these served to inspire Lawrence.

Lawrence added the idea of confining the motion of the accelerating particles by means of a magnetic field. In a vertical field, a horizontally moving charged particle traces out a circle, and the time for one circuit is independent of the particle's velocity. The higher the velocity, the larger the radius of the circular orbit, and the extra length of the path compensates exactly for the higher speed. Thus, the particle could be given a "kick" by a radio frequency (rf) voltage, the frequency being exactly equal to the frequency of rotation. The radio frequency is simply a voltage across a gap, which alternates positive and negative potential in exact synchrony with the particle's circular path. Each crossing of the gap gives the particle an electric kick. With each kick, the particle would gain energy, and the radius would increase, but the period of orbit would be constant.

A magnetic field acts on a charged particle by steering it into a circular trajectory, the plane of the circle being perpendicular to the (north–south) direction of the magnetic field. The radius of the circular path is related to the momentum of the particle (the higher the momentum, the harder the particle is to deflect; thus, the radius is large if the momentum is large) and to the strength of the magnet (the stronger the magnet, the smaller the radius).

Lawrence's first model, built with his student Stanley Livingston, had a magnet whose circular pole pieces were a few inches in diameter;

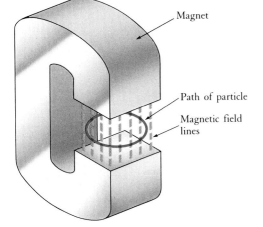

Within a magnetic field, charged particles move in circular orbits whose plane is perpendicular to the magnetic lines of force. The radius of the circle is related to the strength of the field and the momentum of the particle.

The first Lawrence cyclotron. The copper box is installed between the pole pieces of an electromagnet. A small tube leads a stream of hydrogen gas to the center of the box, where a spark strips away the electron from each atom. The protons drift out and are accelerated across the gap between two copper "D's." As the protons gain energy, the radius of their orbit increases, and they gradually spiral out to the limit of the magnetic pole area. Air is pumped from the entire region so that collisions with air molecules do not interfere with the accelerator.

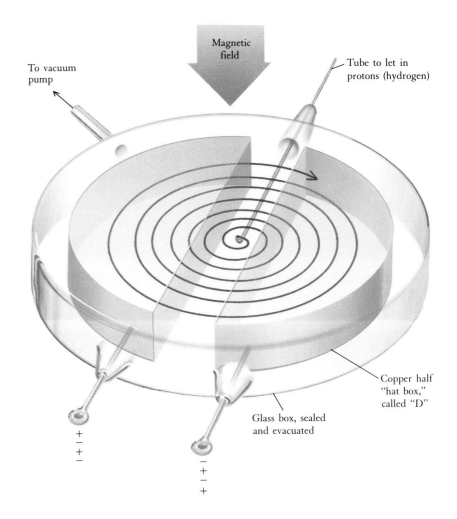

Magnetic field

To vacuum pump

Tube to let in protons (hydrogen)

Copper half "hat box," called "D"

Glass box, sealed and evacuated

+
+
−

−
+
+

in it protons were accelerated to 80 keV. His second model went to 1.2 MeV, and the particles produced nuclear disintegration. The applied radio-frequency voltage was only 1,000 volts, but the protons made more than a thousand circuits as they spiraled out to the rim of the pole pieces. Lawrence was a man of unclouded vision, and so his third machine had 27-inch pole pieces. He called his new machines *cyclotrons*.

Lawrence's machines worked by luck. A condition known as *focusing* was necessary to keep the particles from dispersing away from the neat circular orbits. Any slight deviation from perfect aim would gradually have grown over hundreds and thousands of circuits of the beam. The necessary focusing was automatically provided by the fact that the

Top: An early E. O. Lawrence cyclotron with a diameter of about 4 inches. The energy was just about 1 MeV. *Bottom:* The 11-inch accelerating chamber of the 37-inch cyclotron built in 1937 at the Lawrence Berkeley Laboratory.

magnetic field was weaker outside the orbit, near the rim of the magnet. Mathematical analysis by another young student, R. R. Wilson, showed that this weakening of the magnetic field (later referred to as a magnetic-field gradient) provided a restoring force on particles that deviated from the ideal orbit.

SYNCHROCYCLOTRONS AND SYNCHROTRONS

Analyzing the Lawrence cyclotron, the theoretical physicist Hans Bethe (1906–) noted that when the mass of the accelerated particle began to increase as described by the theory of relativity, the period of the orbit would change, and it would fall out of synchrony with the radio-frequency kicks that accelerate it. This would happen when the velocity of the protons increased to within a few percent of the velocity of light; this set an upper limit on the energy to which particles could be accelerated. The obstacle raised by Bethe gave rise to a new invention, the *synchrocyclotron,* in which the frequency of the applied radio-frequency kick is changed to keep track of the increasing mass. After World War II, synchrocyclotrons of various energies were constructed. These operated in the general domain of a few hundred MeV, the most powerful being a 600-MeV giant. To accommodate these energies, huge magnets were required, with pole diameters ranging from approximately 60 to 500 centimeters. Since the orbits spiral out from radii of 1 centimeter near the center of the magnet to nearly the outer radius, the magnets are solid iron and tend to weigh thousands of tons. The pole pieces had to be machined to special tolerances to provide the required focusing.

A quite different approach was needed for even higher energies. The solution was the *synchrotron.* Here the orbit radius is kept constant, and the magnetic field is increased in synchrony with the gain in momentum. The radio frequency is also changed, since now the orbits are completed in shorter and shorter intervals until the particle's velocity gets close to the velocity of light. As soon as this happens, the frequency can be constant. Since the orbit is fixed, a magnetic field is needed only around the orbit; so a much smaller volume of magnet is required. The disadvantage of the synchrotron is that particles cannot be continuously accelerated; the experimenter accelerates a "batch" to a maximum energy, limited by the maximum strength of the magnets. The batch is used, and the entire cycle is repeated.

The Bevatron, a 6-GeV accelerator that began operating at Lawrence Berkeley Laboratory in 1954. Its most famous discovery was the antiproton in 1955. This accelerator is still operating in 1989.

Magnets are made of special iron (of low carbon content) in order to maximize their magnetic properties. The iron is magnetized by passing electric currents through thick copper conductors attached to the iron. The magnetic field generated by the large currents forces iron atoms into alignment, greatly enhancing the magnetic field between the poles of the iron magnet. This magnetic enhancement reaches a maximum value when all the atoms are aligned. A maximum practical magnetic field based upon iron magnets is between 1.5 and 2.0 Tesla. The Earth's magnetic field is only 10^{-4} Tesla.

The synchrocyclotrons and betatrons of the 1950s were built on the belief that energies in the hundreds of MeV would illuminate the puzzle of nuclear forces. The very productive Nevis accelerator of Columbia University had been upgraded to 400 MeV during its design in 1948 because a theorist speculated that perhaps the muons that had been discovered in the cosmic rays could be produced only in pairs, which would take twice as much energy. Although the theory was wrong, the increased energy turned out to be priceless for producing the heavier pion. Proton and electron synchrotrons built in the 1950s reached the billion-electron-volt range, and started the modern era of large accelerators, rivaling the power of the cosmic rays. Two important proton synchrotrons were the 3-GeV Cosmotron, completed at the Brookhaven Laboratory in Long Island in 1953; and the Bevatron, a 6-GeV giant completed in 1955 at E. O. Lawrence Laboratory in Berkeley, California.

DISCOVERY OF STRONG FOCUSING

In circular accelerators, the particles revolve for tens of thousands of turns during the acceleration cycle; hence stability of the particle orbits is crucial. Any slight deviation of a particle from perfect circularity, for example, a small nudge due to a collision with some residual gas atom, would grow a bit with each turn and lead to eventual loss of the particle, unless focusing forces are operating. In the Cosmotron and the Bevatron a weakly focusing magnetic force was achieved by a decreasing vertical magnetic field, much as we've already seen operating in the cyclotron.

In 1952, a seminal discovery about magnetic focusing was made at Brookhaven. Curiously, it resulted from a visit by a group from the new European Organization for Nuclear Research (CERN), who were designing a larger version of Brookhaven's 3-GeV Cosmotron—a synchrotron designed to accelerate protons. To prepare for their European visitors, a group of Brookhaven experts (Stanley Livingston, Ernest Courant, Hartland Snyder, and others) gathered to discuss how the forthcoming meeting could be most helpful. Because of their discussions, quite serendipitously, they discovered a new focusing principle, which they called *strong focusing*. They discovered that if the magnetic gradients around the circular track of the synchrotron are adjusted to alternately focus and defocus, the stability of the orbits increases; the net effect is

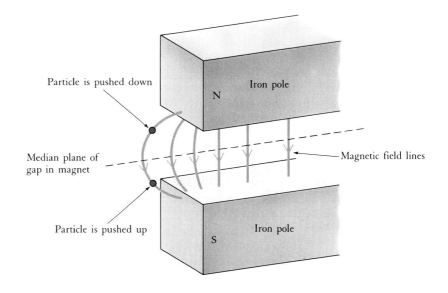

One type of weak focusing is called fringe-field focusing. Curved lines of magnetic force push particles that are above the median plane down and push particles that are below the median plane up.

The radio-frequency quadrupole (RFQ) accelerates protons before injection into the AGS at Brookhaven National Laboratory. The RFQ consists of four electric poles, two positive and two negative, positioned so that identical poles are opposite each other.

focusing. Excited by this surprising result, the physicists then computed that if they increased the strength of the gradients, there would be even greater stability. The variations of the particles from the ideal circular orbit decrease, decreasing the required magnet aperture, and therefore the size and cost of magnets. This discovery led to the building of the alternating gradient synchrotron (AGS).

The AG principle showed that machines could be as large as desired; no limit was seen in the mathematics. Focusing and defocusing could be induced by shaping the pole pieces of the dipole magnets, or by a new invention of the Brookhaven scientists—quadrupole magnetic lenses. Here two north and two south poles are arranged in a way that produces only forces that are perpendicular to the trajectory, pushing the particles toward or away from the axis of motion. Quadrupoles became an essential ingredient in almost all devices that accelerate charged particles.

The CERN visitors were delighted at learning about strong focusing, and returned to Geneva to design a 25-GeV strong-focusing machine, which was completed in 1959. Brookhaven hastened to capitalize on their own discovery and proposed to the U.S. Atomic Energy Commission that they build a 30-GeV machine.

On March 17, 1960, a beam of protons was first injected into the new AGS at Brookhaven, and by July, it had been accelerated to 30 GeV. Although the CERN rival had been completed a year earlier, the AGS was to become the most productive accelerator ever; no other accelerator has made so many important discoveries.

Strong focusing is created by a quadrupole magnetic field that acts as a lens to either focus or defocus the particle paths. The best way to keep the paths stable is to alternately focus and defocus them. One arrangement of magnetic poles defocuses (*left*), whereas the opposite arrangement focuses (*right*).

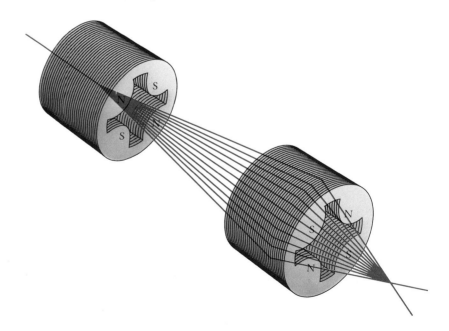

THE DISCOVERY OF
THE MUON NEUTRINO

The first major discovery at the AGS was called the "Two-Neutrino" experiment, carried out in 1961. It was led by three Columbia University professors (noted for their shy and retiring natures, and known to the Brookhaven physicists as "Murder, Incorporated"). They were Jack Steinberger, Melvin Schwartz, and Leon Lederman, who received the Nobel Prize in physics for this experiment in 1988.

The experiment was intended to study the weak force, the only one of the four basic forces that affects neutrinos, and which had, therefore, been extremely difficult to study. A neutrino passing through a piece of matter interacts with another particle or atom so rarely that a single neutrino, if passing through 100 million miles of solid steel, would stand only a 50% chance of being stopped or deflected. Only when immense numbers of neutrinos pass through an object are a few of them likely to collide with something and thereby signal their existence. Small wonder that neutrinos can escape from the cores of stars, an important fact in stellar dynamics, and another example of how neutrinos and their properties are essential for understanding cosmological processes.

It was at a coffee break at the Pupin Physics Building on the Columbia campus that Melvin Schwartz, formerly one of Jack Steinberger's doctoral students, first proposed the possibility of creating beams of high-energy neutrinos in the laboratory for use in research. With Leon Lederman and other colleagues, the group resolved to use the powerful Brookhaven accelerator to create their beam.

Unknown to the Columbia group, the Italian physicist Bruno Pontecervo, working in the Soviet Union, independently proposed that neutrinos would make useful projectiles. However, his proposal missed the importance of using high-energy neutrinos.

The Brookhaven accelerator was rigged to hurl swarms of protons at 15 GeV into a target made of beryllium metal. The immense energy of the resulting proton impacts not only tore beryllium atomic nuclei apart into their constituent protons and neutrons, but also created pions, which are unstable and can decay into muons and neutrinos. Since the pions decay while moving at high energy, the product muons and neutrinos share this energy. All of these particles were swept along in the same direction as the original protons streaming from the accelerator.

To filter out all but the neutrinos, the group erected an obstacle—a 40-foot-thick barrier of steel taken mostly from a scrapped World War II battleship (a modern version of "swords into plowshares"). The steel armor blocked the protons, neutrons, nuclear fragments, undecayed pions, and other flotsam produced by the proton bombardment, allowing

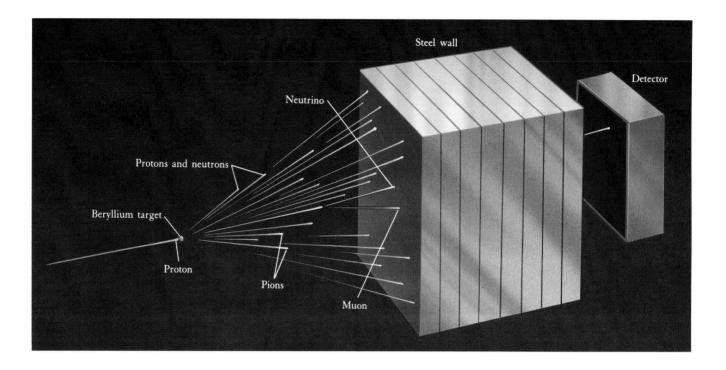

The neutrino experiment at the Brookhaven AGS. Protons from the accelerator strike a beryllium target, producing a variety of subnuclear particles including pions. During a 20-meter flight, some of the pions decay into muons and neutrinos. All the particles smash into a thick steel wall, which absorbs everything but the neutrinos. The neutrinos pass through the wall, and some very small fraction interact in the detector.

only the neutrinos to pass. The result was the world's first research beam of high-energy, high-intensity neutrinos (see the figure on this page). Behind the steel barrier, a spark chamber detector was constructed to observe the extremely rare neutrino collisions. Although only a minute fraction of the neutrinos collided in the detector, the beam contained so many neutrinos that a significant number of them were seen in the 10-ton detector that the Columbia group had built. Despite the extreme rarity of neutrino impacts, some 56 of them were detected in an eight-month period.

Pauli's neutrino is born with an electron in radioactive decay. This means that in collisions Pauli's neutrino can surely produce electrons. The pion decay into a muon and a "neutrino" means that this neutrino must be capable of producing muons in collision. Are these two neutrinos identical? If there were only one kind of neutrino, then the collisions of the neutrinos with the aluminum atoms would have produced equal numbers of electrons and muons. However, only muons were observed. Hence, these impacts were being caused by a new kind of neutrino—the one associated with muons. This discovery was to have far-reaching consequences.

Until this Columbia experiment, the only neutrino known was that produced by the beta-decay process, which also produces an electron. But most of the neutrinos coming through Columbia's steel filter were born in pion decay and were accompanied by muons. These neutrinos were seen in the experiment to be capable of producing muons when they collide—never electrons. They were therefore named *muon neutrinos.* To articulate the difference between these two types of neutrinos, it was suggested that they differed in *flavor:* that the neutrinos that result from beta decay are electron-flavored, and these new ones were muon-flavored. As you might guess, the word *flavor* was not suggested with absolute seriousness, yet the concept of flavor became crucial to enabling the creation of the standard model.

In this generalization, we have quarks of six different flavors, and we have leptons of six flavors. We assign a flavor to each experimentally distinguishable member of a class of particles. This simple idea of flavor exploded into recognition in the dramatic case of the two neutrinos, where the distinction was subtle; its recognition required a Nobel-class experiment.

This pioneering neutrino experiment was followed by many others, with the detectors increasing in weight to several hundred tons. The intensity of neutrino beams has become so great that in 1987 more than one million neutrino events were collected at the Fermilab Tevatron.

GELL-MANN AND THE EIGHTFOLD WAY

By the early 1960s the number of particles discovered by the new accelerators and associated particle detectors had grown to almost one hundred. Since these were mostly the products of strong forces they were collectively called *hadrons,* using the Greek word for *strong.*

In 1962, the world's largest bubble chamber, 80 feet long and containing 900 liters of liquid hydrogen, was exposed to a beam of negatively charged kaons from the AGS. The goal of the experiment was to test a new hypothesis by Murray Gell-Mann, of the California Institute of Technology, that attempted to organize the hadron particles known by 1962.

What Gell-Mann (1929–) had discovered was that many of the particles could be organized into families of eight or ten, with properties that are mathematically the same as those of what is called a "group of eight" (octet) in abstract algebra (a branch of mathematics rarely before applied to physics!). Groups of ten (decuplets) were also discovered by a clever arrangement of the particles. Gell-Mann whimsically titled his discovery using the Buddhist term "The Eightfold Way." (This symmetry was also discovered, simultaneously and independently, by Yuval Ne'eman, an Israeli theorist working in England.)

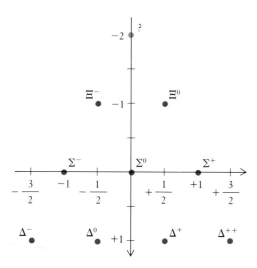

The baryon decuplet. Nine known particles are organized according to their individual quantum properties. The vertical and horizontal scales are related to quantum numbers that define the properties of strangeness and isospin. The missing state at the apex suggested the existence of a particle with well-defined quantum numbers. When discovered, it was named "omega minus."

The key to testing Gell-Mann's hypothesis was that it predicted the existence of a new particle with particularly exotic properties. Nine known particles, because of their symmetric properties, could be arranged in a pyramidal structure, shown in the margin. The particles were graphed in accordance with certain quantum numbers with exotic names such as "isotopic spin" and "strangeness." Later we will see that strangeness was related to the properties of the quarks out of which they were all made, and isotopic spin was related to the electric charge. Symmetry screamed out for the existence of a tenth particle, which had already been named Ω^- (omega minus). The goal of the Brookhaven bubble-chamber experiment, led by Nicholas Samios, was to find the missing particle.

In December 1963, the experiment started. Every few seconds, a burst of negative kaons, perhaps between ten and twenty, would tear through the just-sensitized monster bubble chamber, lights flashed, and three immense cameras moved their 70-millimeter film. The kaons would collide with the protons (nuclei of the liquid hydrogen) and each collision would produce a spray of particles. As the day ended, the pictures were developed and quickly scanned for something that would indicate an entity with the predicted characteristics of charge minus one and strangeness minus three.

The photo on the facing page was the 50,321st picture—photographic proof of the existence of the Ω^-. (Twenty-four years later, a 1987 experiment at Fermilab's Tevatron acquired 100,000 Ω^- particles in order to measure the magnetism of this once-exotic particle!) Samios eventually became director of the Brookhaven Laboratory. Gell-Mann's organization was confirmed, but what did it mean?

THE QUARK HYPOTHESIS

Gell-Mann's "Eightfold Way" organized the new particles into family groups that reminded one of the periodic table. Recall that an understanding of the table came only after it was known that atoms were made of electrons and nuclei. This grouping of hadrons suggested to some physicists that there could exist underlying "fundamental" particles that might comprise the observed particle zoo. In 1964, Murray Gell-Mann and another physicist, George Zweig (also from Cal. Tech.), independently proposed an underlying particle structure. Both men proposed a triplet of fundamental particles that would make all the hadrons.

Gell-Mann was giving a series of lectures at Columbia University while working on this problem. In response to a question by Columbia's

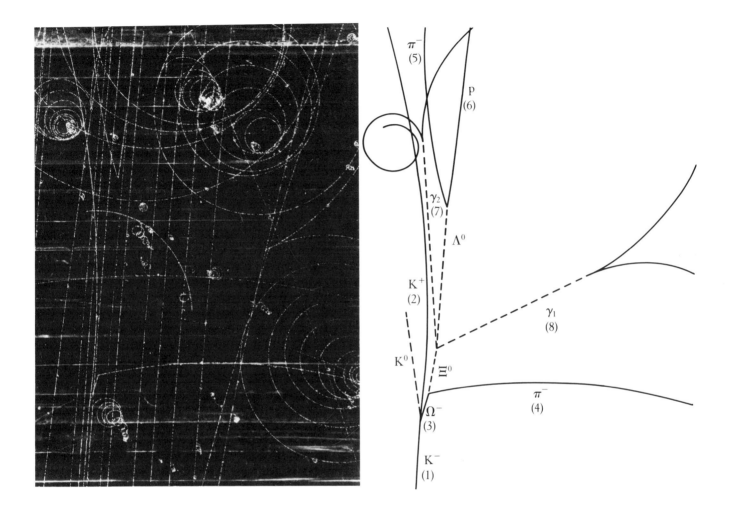

Bubble chamber photograph revealing the existence of the Ω^- particle. The 7-foot hydrogen-filled chamber was exposed to a K^- beam at Brookhaven National Laboratory. A K^- particle collided with a positron at (3). The resulting spray of particles included a negatively charged omega meson, Ω^-, whose existence had been predicted by theoretical physicists. The sketch at the right identifies the particles that resulted from the interaction, including the Ω^-, labeled (3).

Robert Serber, Gell-Mann admitted that an underlying triplet of particles could give rise to his groups of eight or ten known particles, but would have rather "quirky" properties. For example, each of the three would have to have a fractional charge, of plus or minus one-third or two-thirds of the electron charge. They would also each have to carry one-third of the properties of a proton or neutron. By the time the lectures were published, Gell-Mann had concluded that the triplet idea was attractive in spite of the bizarre properties of the objects. He had also done some of the literary and linguistic manipulations for which he is famous and labeled his underlying particles *quarks*, referring to a line from James Joyce's *Finnegans Wake,* "Three quarks for Muster Mark." *Quark* is the German word for a cottage-cheese sort of dessert, and is also

German slang meaning "nonsense." Gell-Mann's word stuck, and Zwieg's name for the particles, "aces," was rapidly lost.

Gell-Mann's quark proposal introduced a new simplicity and order into the particle zoo. All known hadrons could be made from combinations of three basic kinds of quarks. Eventually each kind of quark became known as a flavor, and the initial three flavors are now known as "up," "down," and "strange." The up quark has an electric charge of $+\frac{2}{3}$ unit, the down quark has $-\frac{1}{3}$, and the strange quark, $-\frac{1}{3}$. Nuclear matter, known as *baryonic matter* (e.g., the neutron or the proton), is made out of three quarks (two ups and a down for the proton, two downs and an up for the neutron). (The Ω^- is made out of three strange quarks.) *Mesons,* which were originally proposed as exchange particles that hold nucleons together, are made from a quark–antiquark pair. For example, a pion is made from an up and an "antidown," or a down and an "antiup." The strange mesons, such as the kaons, are made by including strange quarks with the ups and downs in quark–antiquark pairs. For example, the K$^+$ consists of an up quark and an "antistrange" quark: u$\bar{s}$. (See the table for more details on baryon and meson composition.)

It is important to point out that the hypothesis for the existence of quarks was motivated by the existence of the baryons and mesons that arranged themselves into these "Eightfold-Way" patterns. That three quarks with cleverly adjusted properties could replicate all the known

COMPOSITION OF BARYONS AND MESONS

	Particle	Quark composition	Charge
Baryons (qqq)			
p	proton	uud	$+\frac{2}{3} + \frac{2}{3} - \frac{1}{3} = +1$
n	neutron	udd	$+\frac{2}{3} - \frac{1}{3} - \frac{1}{3} = 0$
Λ	lambda	uds	$+\frac{2}{3} - \frac{1}{3} - \frac{1}{3} = 0$
Λ_c	lambda "c"	udc	$+\frac{2}{3} - \frac{1}{3} + \frac{2}{3} = +1$
Ω^-	omega minus	sss	$-\frac{1}{3} - \frac{1}{3} - \frac{1}{3} = -1$
Mesons (q$\bar{\text{q}}$)			
π^+		u$\bar{\text{d}}$	$+\frac{2}{3} + \frac{1}{3} = +1$
π^-		$\bar{\text{u}}$d	$-\frac{2}{3} - \frac{1}{3} = -1$
π^0		$\dfrac{\text{u}\bar{\text{u}} + \text{d}\bar{\text{d}}}{\sqrt{2}}$	$(+\frac{2}{3} - \frac{2}{3}) + (-\frac{1}{3} + \frac{1}{3}) = 0$
K$^+$		u$\bar{\text{s}}$	$+\frac{2}{3} + \frac{1}{3} = +1$
K$^-$		$\bar{\text{u}}$s	$-\frac{2}{3} - \frac{1}{3} = -1$

baryons and mesons very much favored the quark hypothesis. But harder evidence was needed before this would be widely accepted. Are quarks *real* or just a mathematical convenience?

EXPERIMENTAL VERIFICATION
OF QUARKS

A proton can be accelerated in a machine because it is a stable particle. So is an electron, but because of its relatively low mass, an electron in a circular orbit radiates away vastly more energy than a proton does. That energy must be continually restored by the system, and this fact limits the maximum useful energy of circular electron accelerators.

It was to get around this problem that the Stanford Linear Accelerator Center (SLAC) was built in 1961. The 3-kilometer-long series of radio-frequency cavities pushed the electron beams in a straight line up to 20 GeV. (Subsequently the accelerator structures were improved until, in 1987, SLAC produced 50-GeV electrons.)

An aerial view of the Stanford Linear Accelerator Center. The 3-kilometer-long linear accelerator house can be seen crossing under Highway 82. Experiments are carried out at the end stations in the foreground and in an underground storage ring.

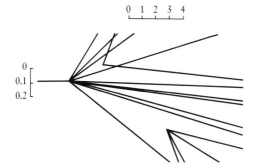

Evidence for the existence of quarks. The collision at the left occurred in a beam of 800 GeV protons injected into a Fermilab bubble chamber; the important products of the collision are an invisible neutral "charmed" meson ($D^{\underline{0}}$) and its also invisible anticharmed twin ($\overline{D^0}$). The D^0 decays into two charged particles after traveling about 0.3 millimeters. The $\overline{D^0}$ flies 0.8 millimeters before decaying into four charged particles (bottom right). The lifetime of the D^0 meson, composed of a charmed and an up quark, is about 10^{-12} seconds. This event was photographed using a special holographic technique that improves the spatial resolution.

A series of experiments carried out at SLAC during the late 1960s and early 1970s proved that quarks are not just purely mathematical entities, but real constituents of protons and neutrons. The SLAC experiments involved bouncing high-energy electrons off protons. The experimental team was headed by Henry Kendall and Jerry Friedman, two professors from M.I.T. In many ways they were carrying out the modern equivalent of the Rutherford scattering experiment, but at much higher energies. Thus, instead of looking at the structure of atoms and discovering the nucleus, these experimenters were looking at the structure of the proton and, as it turned out, finding quarks. An important role in the SLAC experiments was played by two leading theoretical physicists, James "B.J." Bjorken, then at SLAC, and Richard Feynman, the Nobel laureate from Cal. Tech. Bjorken had noted that the scattering behavior should increase with energy and momentum, depending on how many sources of scattering there are inside the proton. However, initially Bjorken's analysis was not understandable to the experimenters. Feynman came to visit, saw what was being measured, and listened to Bjorken's formalism. That night, he awakened at three in the morning with his own synthesis of the experimental results and the expected theoretical behavior put into a new, very picturesque, description of what was happening. He saw that the experimental data implied that the proton was indeed made up of smaller pieces that scattered the electrons. Feynman referred to these pieces as *partons* (particles that were parts of the proton). Why did he and the M.I.T./SLAC people refer to these as partons rather than quarks? Probably because they did not want to prejudice their results towards any specific theory. Also, Feynman was convinced that other objects accompanied quarks inside the proton. We will see later that these are gluons—the basic carriers of the quark–quark strong force. It was convenient to have a collective name, hence, partons.

As the electron-scattering experiments continued, it became clear that partons behaved similarly to pointlike particles making up each proton. By the early 1970s, many additional experiments carried out at ever-higher energies confirmed and refined the SLAC results. Electron scattering was supplemented by higher-energy muon scattering and neutrino scattering at CERN and Fermilab. The last resistance to the quark idea collapsed in the "November Revolution" of 1974 as we shall see. Quarks with all their curious properties existed. One of these curious properties is their pointlike nature. As point particles, they seemed indivisible and thus truly fundamental. Also, as point particles with no apparent volume, there would be no problem squeezing huge numbers of them into small volumes, as would occur in a black hole or in the big bang itself. The pointlike nature of fundamental particles is important for the high densities of the early universe, since finite-sized composite particles like the proton could not be packed as densely as the big-bang theory requires and remain separate entities.

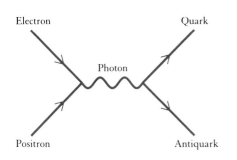

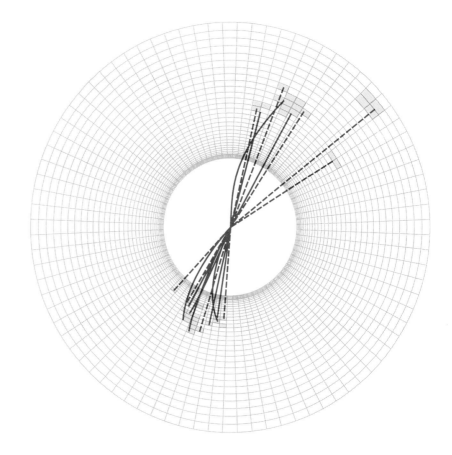

A qualitative demonstration of the existence of quarks. A collision of an electron and a positron (moving in a path perpendicular to the plane of the paper) results in two narrow sprays of particles, called "jets." A satisfactory explanation for why this pattern is so frequently seen is that the electron–positron annihilation produces a "virtual" photon, which in turn dissolves into a quark and an antiquark flying in opposite directions. These quarks change to jets as they leave the scene of collision. This event was recorded at the PETRA accelerator in Hamburg, Germany. The paths of the particles were reconstructed by computer from ionization tracks and from the pattern of energy (shown in color) deposited as the particles struck the inner layer of the 2.4-meter-long cylinder of the detector.

Within a few years it became clear that Feynman's partons were indeed quarks and gluons (which we will meet in the next section). Quarks were not just mathematical entities denoting properties of symmetry, but real particles capable of deflecting electrons, even though they were apparently confined inside neutrons and protons.

"Confinement" was another curious property of quarks. It was invented by theorists after the failure of countless attempts to observe free quarks with a one-third integral electric charge. Quarks, they said, were permanently confined inside the hadrons and it was no more possible to isolate a single quark than to separate the north and south poles of a bar magnet. We will see later that confinement eventually taught us something very deep about the nature of the quark–quark force.

Although the November Revolution crushed the last resistance to the quark hypothesis, it is fair to say that the reality of quarks evolved gradually out of a long series of experiments.

Bubble chamber photograph of a high-energy gamma ray (photon) colliding with a hydrogen nucleus. The collision produces one energetic negative electron and one negative-electron/positive-electron pair as well as a new gamma ray which converts, by subsequent collision, into another electron–positron pair. The strong magnetic field curves the trajectories of positive and negative particles in opposite directions.

THE DISCOVERY OF COLOR CHARGE: ELECTRON STORAGE RINGS

In the 1960s, electron aficionados had devised electron–positron storage rings, where separate bunches of oppositely charged electrons and positrons circulated in the same vacuum pipe around a magnetic ring, moving in opposite directions. Conveniently, both beams are focused by the same quadrupole magnets in the ring. The bunches meet at selected locations, and for the most part pass through one another like two diffuse swarms of bees heading in opposite directions. Occasionally there is an encounter as improbable as two machine-gun bullets colliding in midair. The collision is very violent—the total energy of both particles is invested in the products of the collision. The annihilation of the colliding bits of matter and antimatter dissolves into a wide variety of possible final states: photons, pions, kaons, protons, antiprotons, and so forth. As the technique matured, the density of the bunches increased, and the collision rates improved. Two of the early electron–positron colliders were the ADONE machine near Rome and the Stanford Positron–Electron Annihilation Rings (SPEAR).

The earliest success of these studies was again to refine our understanding of quarks. The simplest experiment was to count the average number of baryons and mesons emerging from these electron-positron collisions. These were presumably made out of quarks materialized in the annihilation. Results from the ADONE machine in Italy indicated (in 1971) that the yield of pions was three times larger than the simple quark theory predicted. What was happening?

Murray Gell-Mann suggested a new quantum number, analogous to electric charge but connected to the strong force. To account for the ADONE data, the new charge had to have *three* states, as contrasted with the two states of electric charge. We use "plus" and "minus" to distinguish the two states of electric charge. Gell-Mann chose *color* as the word to distinguish the three states of the quark strong charge, which is now called *color charge*. So quarks come in three "colors," say, red, blue, and green and each color of quark can produce a pion along an independent pathway in the ADONE collisions. This fact triples the probability that a pion will be produced.

Gell-Mann's color concept was accepted fairly quickly, because it also explained several other puzzles about the quark idea. For example, quarks combine *only* in the forms q$\bar{\text{q}}$ (meson) or qqq (baryon) because these are the only combinations in which the total color charge cancels, leaving no net color for the ordinary known particles (see the figure on the facing page). (The addition of the three colors leads to white, i.e., the absence of color; so this made the analogy to real color more interest-

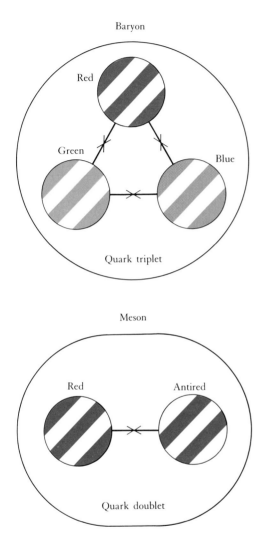

The baryons such as protons and neutrons are made of three quarks, each having one of three different colors. The mesons are made of a quark of one color and an antiquark having the anticolor of that color.

ing.) Antiparticles have an anticolor that cancels the color of particles, again leading to colorless mesons. The color concept also explained the apparent contradiction between the three-quark baryon structure and the Pauli exclusion principle: The "uud" composition of the proton would have two fermions in the same quantum state, an impossibility in quantum physics, unless the two up quarks have different colors. The strong force between quarks is then due to the color property. [The strong force has been given another name suggested by the color quantum number—quantum chromodynamics (QCD).] In the continuous sonata between theory and experiment, the next movement belonged to experiment.

NEW PARTICLES AND NEW QUARKS

The electron–positron machines scored yet another major success, this time at Stanford's SPEAR. In what has since been called the November Revolution, physicists using SPEAR made a seminal discovery, one which firmly established the quark theory. The leader of the SLAC group was Burton Richter, who had played an important role in building the SPEAR machine in the period between 1970 and 1973. At the same time, Richter's group built a very complex particle detector to observe the head-on electron–positron collisions, and began observations in 1973. The experimental team included Roy Schwitters, Martin Breidenbach, and an experienced ex-bubble-chamber team from nearby Lawrence Berkeley Laboratory, led by George Trilling and Gerson Goldhaber.

Results of these collisions in the energy interval from 2.5 to 4.0 GeV were mystifying, and the ever-watchful and recently converted quark theorists began to fear for their theory. The SPEAR accelerator could tune the total energy of the colliding particles very precisely. One procedure for studying the reaction rate was to vary this energy to see whether the number of collisions depended upon the energy.

On November 9 and 10, 1974, the group had decided to repeat the measurements very carefully to explore some anomaly near 3.1 GeV (each beam had 1.55 GeV). The repetition, in steps of 0.001 GeV, turned up one of the more spectacular discoveries of the decade. The number of collisions increased one-hundred-fold between 3.100 and 3.105, then dropped very rapidly by energy 3.120. This clearly signified a new particle with mass 3.105 GeV. Previous searches had missed this sharp spike.

What was remarkable was the narrowness of the counting peak, or mass. In quantum mechanics, a particle does not have a precisely defined mass. There is an associated uncertainty that is related to its lifetime.

According to Heisenberg's uncertainty relations, the longer a particle's lifetime, the better defined its mass. Electrons, for example, are stable and have a precise mass, whereas the Δ^{++} particle, a bound state of a positive pion and proton, lives for only 10^{-23} seconds, and has a mass uncertainty (or mass width) of tens of MeV. Now ordinarily, the heavier the particle, the more states there are into which it can decay easily, and hence the shorter its lifetime. The new particle found at SLAC (it was called psi, ψ) is three times heavier than a proton, and so it should have had a very short lifetime. Instead its narrow mass width suggested a very long lifetime. What prevents it from decaying? The answer explained the excitement of the November Revolution: Only if the new object was made of a new form of matter could it not decay into particles made of old (u, d, s) quarks. This new form of matter was in fact most elegantly attributed to a new quark, which was named "charm."

Part of the excitement at the time was the fact that the same discovery was made simultaneously at the Brookhaven Laboratory, 3,000 miles away. Samuel C. C. Ting of M.I.T. was studying the production of electron–positron pairs that occur when a proton strikes another nucleon. Ting's experiment was called "inclusive," because it selected one thing (in this case, an electron–positron pair) to study and ignored everything else that might come out of the collision. The idea was to measure everything about the emerging positive and negative electrons. The electron and positron can be thought of as the decay product of a temporary (or virtual) photon:

$$\text{``}\gamma\text{''} \rightarrow e^+ + e^-$$

In quantum mechanics a particle such as a photon can "borrow" a certain amount of energy from the vacuum, provided it keeps it for a short time—the more energy borrowed the shorter the time. During the time the photon has this energy, it is known as "virtual." The virtual photon can have any mass: the higher the mass, the smaller the region of space from which it emerges after the nuclear matter is heated by the collision. The nice thing about this reaction is that by measuring everything about the electron–positron pair, you can determine the virtual mass. This type of inclusive experiment had been invented by a Columbia University group in 1968, where the emerging pairs were traced out over a mass range of 1 to 6 GeV. The data were later interpreted by Sidney Drell and Tung Mo Yan as the annihilation of a quark in the proton and an antiquark in the target to give birth to the virtual photon. The process had been widely studied since then to learn about the properties of bound quarks, and its agreement with electron, muon, and neutrino scattering was a very strong factor in establishing the consistency of the quark theory.

The Columbia data had also shown a suspicious but inconclusive bump, or shoulder, near 3 GeV. Six years later the technology of particle

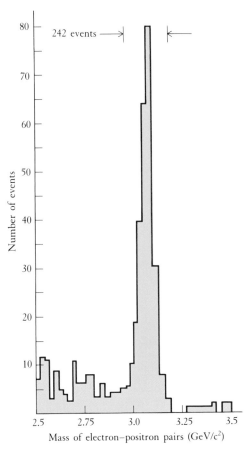

Ting's discovery of the J particle. For each electron–positron pair he recorded, researchers calculated the mass of the pair, which gave the mass of the parent particle. The existence of a new particle is indicated by the narrow peak at 3.1 GeV. The particle was simultaneously confirmed at Stanford by observing the reverse reaction, where electron–positron collisions produced the Ψ. It was eventually shown that the J/Ψ is a bound state of a charm quark and an anticharm quark.

detection had advanced considerably. Multiwire proportional wire chambers (MWPCs) enabled measurement of particles with ease and precision. Ting's experimental apparatus was replete with layers of MWPCs, including a strong magnet, toward which the electrons were directed. The magnet deflected positive particles to the right arm, negatives to the left arm. MWPCs recorded the paths of particles and, therefore, the deflection induced by the magnet. This gave both the momenta and the angle between the electron and the positron. The next pieces of apparatus were scintillation counters to give fast electronic-pulse information about the passage of the electrons, and then special counters to establish that the particles were indeed electrons.

From the measurements, the computer can calculate the mass of the virtual photon and graph the number of events at each mass. The graph on this page shows Ting's Nobel Prize–winning result: At 3.1 GeV the virtual photons were being supplemented by a new particle that Ting called J. (The character J in some Chinese dialects is pronounced "Ting.") In order not to show favoritism between the East and West coasts, the particle is now called the J/ψ.

Ting's success with the Brookhaven J experiment unleashed him as a new force on the scientific world. Between 1982 and 1989, Ting supervised the construction of the most complex detector ever assembled to carry out an experiment at CERN's LEP accelerator. More than 400 scientists from thirteen nations assembled the 10,000-ton "L3" detector in a giant cavern underneath the Jura Mountains on the Swiss–French border. Data collection is scheduled to begin in July of 1989.

Although the discovery of charm was a major experimental feat, charm had in fact been predicted by theorists J. D. Bjorken and Sheldon Glashow as early as 1964. At that time, a triplet of quarks were known: u, d, s. The two-neutrino experiment had established a *four*-lepton pattern: (e, ν_e) and (μ, ν_μ). Bjorken and Glashow believed there should be a symmetry between quarks and leptons and that such a symmetry required a fourth quark and a corresponding pattern (u, d) and (c, s). The idea was charming and they named the new quark *charm,* designated c. The discovery of charm in 1974–1975 established these suggested patterns and the form of the standard model. In this viewpoint, quarks and leptons come in pairs. The quarks (u, d) and the leptons (e, ν_e) belong to the first (and lowest mass) generation. The second generation has (c, s) and (μ, ν_μ).

It took about a year to prove conclusively that the J/ψ particle was composed of a new quark. In fact, the composition is a bound state of the charm quark and its antiparticle c$\bar{\text{c}}$, analogous to the u$\bar{\text{u}}$ (up, antiup) and u$\bar{\text{d}}$ (up, antidown) objects we call mesons. Confirmation came in 1975 with the identification of particles with quark combinations c$\bar{\text{u}}$ and c$\bar{\text{d}}$. These "charmed mesons" had just the properties the quark theory had predicted.

In 1978, at Fermilab's 400-GeV proton machine, in an experiment of the Drell–Yan type, yet another new quark was discovered. The experimental team was from Columbia, Stony Brook, and Fermilab (CSF). The data showed an enhancement, in fact, three closely spaced bumps, in a continuum of pairs of muons from virtual photons, like the original Columbia research at Brookhaven in 1968. The enhancements represented a new particle called Υ (upsilon), which was again very narrow, replaying the J/ψ story, but now at 10 GeV of mass. When the dust settled, the $\underline{\Upsilon}$ turned out to be a bound state of a new quark and its antiquark $\overline{b}b$ (b stands for "beauty" or "bottom"). The additional peaks represented excited states of the $\overline{b}b$ atom. Again theorists rushed in to claim that "b" implies "t," that is, if beauty, then truth: A partner must exist to fill out the third generation.

The experiment that found upsilon had what was becoming a standard array of MWPCs, magnets, scintillation counters, and an elaborate electronic data-acquisition system, which selectively recorded interesting events onto magnetic tape. All this equipment represented a breakthrough in being able to handle enormous quantities of information coming in at a rate of hundreds of events a microsecond. Precision in the measurement of the masses of muon pairs was another advantage of this system.

Once beauty was discovered, it was clear that there were more than three flavors of quarks. In fact with this discovery, the natural question was "How many flavors are there?" In particular, given higher and higher energies in accelerator experiments, would we continually find new quarks, and end up with yet another particle zoo? As we will see in the next chapter, here cosmology came to the rescue of particle physics, saying firmly that the number cannot increase indefinitely.

In 1976, in the very productive SPEAR collider, a new lepton was discovered, heavier than the muon but otherwise very similar in properties. Stanford physicist Martin Perl sorted this out of SPEAR data, and eventually convinced his colleagues at SLAC and throughout the rest of the world that he had discovered the "tau" lepton. Now we see the power and authority of the theoretical idea of symmetry. From the table on the facing page we see that tau belongs to the third generation and theorists immediately assumed that, analogous to the electron and its electron neutrino, and the muon and its muon neutrino, there must be a tau neutrino to accompany the tau.

By the end of the 1970s, all ingredients of the standard model were in place. We had six quarks (of which the truth, or top, quark had not been found), we had six leptons (of which the tau neutrino was still missing), and we had a logic for how to organize them into the three generations indicated in the table. Experiments had established the existence and properties of the leptons and of the c and b quarks. Note a curious symmetry of the generation structure: The sum of the charges in

PARTICLES IN THE STANDARD MODEL

Generation	Leptons				Quarks			
	Particle name	Symbol	Mass at rest (MeV)	Electric charge	Particle name	Symbol	Mass at rest (MeV)	Electric charge
I	Electron neutrino	ν_e	About 0	0	Up	u	About 5	$+\frac{2}{3}$
	Electron	e^-	0.511	-1	Down	d	About 7	$-\frac{1}{3}$
II	Muon neutrino	ν_μ	About 0	0	Charm	c	1,500	$+\frac{2}{3}$
	Muon	μ^-	105.7	-1	Strange	s	About 150	$-\frac{1}{3}$
III	Tau neutrino	ν_τ	Less than 35	0	Top/truth	t	>41,000	$+\frac{2}{3}$
	Tau	τ^-	1,784	-1	Bottom/beauty	b	About 5,000	$-\frac{1}{3}$

each generation is zero. In the first generation, three colored up quarks times $+\frac{2}{3} = +2$, three down quarks times $-\frac{1}{3} = -1$, the electron has -1, and the neutrino has charge $= 0$. Thus the generation structure indicates a deep connection between quarks and leptons. Each generation is a heavier Xerox copy of the previous generation.

GLUONS AND THE STRONG FORCE

Fundamental forces or interactions occur via the exchange of bosons. The exchange particle for the electromagnetic interaction is the photon (light). In the old Yukawa picture (discussed in Chapter 3), the meson was thought to be the exchange particle between protons and neutrons. With the quark model, protons, neutrons, *and* mesons are made of quarks that are strongly bound together. We thus learn that the primary strong force is between quarks. The role of mesons is a kind of secondary, and much more complex, example of the strong force. The exchange particle between quarks and the true carrier of the strong force is the *gluon*. The properties of the gluon came out of the standard-model theory. Recall that the photon played an analogous role for the electromagnetic force.

Evidence for gluons came in 1978 from an electron-positron machine at Hamburg in West Germany. The machine, called PETRA, was able, like its Stanford twin PEP, to observe collisions up to 30 GeV, and here in the pattern of produced particles, the gluon was read.

This collision recorded at PETRA is one example of the three-jet structure of many electron–positron annihilation collisions, presented as evidence for the existence of gluons. In some fraction of two-jet events such as the one illustrated on page 105, a quark should radiate away a gluon. The gluon would appear as a group of pions, kaons, and other particles somewhat broader than the quark jet. The observations are very much what is expected.

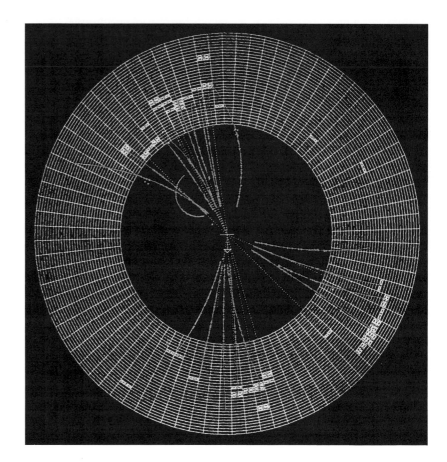

Gluons and photons share several properties as force carriers. Both are bosons of spin one; both are massless. However, gluons differ from photons in that they carry the strong (color) charge, whereas photons do not carry any electromagnetic charge. This difference was crucial in producing a curious feature of the quark–quark force, first recognized by David Politzer, then a graduate student at Harvard, and David Gross, a new professor at Princeton, and his student Frank Wilczek. The force between quarks *increases* as the quarks separate. In contrast, the force between electrons decreases with separation. Since the strong force is carried by gluons, to say that the force is stronger is to say that there are more gluons. The farther apart two quarks are, the more force-carrying gluons there are between them, and the *stronger* the force. Such a mechanism means it would take infinite energy to completely separate two quarks from each other. Thus, color explains the process of confinement, that is, the impossibility of producing a free quark. On the other hand, when three quarks or a quark–antiquark pair are *close* to each other, the minimum number of gluons are exchanged, and the force is minimized.

This is referred to as *asymptotic freedom,* meaning that at close distances, the quarks behave like free particles. Hence, quarks may never have escaped from inside baryons and mesons at any time other than at the big bang, when they were all squeezed together so tightly that the whole universe was a "quark soup."

The color structure tells us also about the properties of the gluons. Since they are absorbed and emitted by quarks, they can change the color of the quarks; that is, a red–blue gluon changes a red quark to a blue quark, and so forth. There are also red–red, blue–blue, and green–green gluons, so that there are nine possible gluon states in all (although mathematically only eight of them are independent). Thus we see a kind of pattern: The electromagnetic force requires one photon; the weak force requires three carriers, W^+, W^-, and Z^0; and the strong force requires eight gluons, each labeled by two colors. (The gluons actually carry one color and one anticolor e.g., red/antiblue.) The properties of the weak force indicated that the weak force carriers were massive, but no mass scale could be calculated reliably.

Bit by bit, the loose ends of the standard model became fixed, as experiment suggested theory and theory suggested experiment. The culmination of this duet came to a theoretical head during the 1970s, and hit an experimental crescendo in the 1980s; it goes by the name of "electroweak symmetry breaking" and led to the unification of the electromagnetic and weak forces and to the prediction of precise masses for the weak-force carriers.

ELECTROWEAK UNIFICATION

The scientific problem that captured Albert Einstein's attention after his work on general relativity in 1916 and until his death in 1955 was the quest for a "unified field theory." Einstein failed! He never found this "holy grail" of physics. However, in recent years, physicists have succeeded in unifying at least two out of the four forces of nature, and this unification gives powerful hints as to how the remaining two forces might be included in the unification scheme.

To further illustrate what we mean by unification, let us describe how the two forces, the nuclear weak force and the electromagnetic force, were unified. At first glance, these might appear poor candidates for merger. The electromagnetic force has infinite range, with its intensity falling off as the inverse of the square of the distance between electric charges. The nuclear weak force appears to interact with all fermions, that is, all quarks and leptons, all of which have a spin of $\frac{1}{2}$ unit. This force has such a short range that it is not effective outside the atomic nucleus. It can convert one flavor of quark to another, and thus can

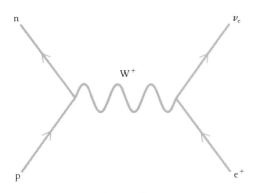

The radioactive decay of a neutron, an example of a weak interaction mediated by a charged boson. A neutron n emits a W boson and is converted to a proton p. The W^+ decays to an electron (e^+) and an antineutrino (ν_e).

transform neutrons to protons or vice versa. Thus, it is responsible for the many nuclear decays we collectively call radioactivity.

In Chapter 3, we introduced photons as the carriers of the electromagnetic force. We also learned that gluons are the carriers of the strong force. But in the Fermi theory of the weak force, what are the force carriers? Fermi avoided the issue completely by proposing an oversimplified theory that worked very well in the low-energy experiments of the 1930s. By the 1950s, however, various theorists had corrected Fermi's theory so that it could be applied to higher-energy processes. Here, the idea of force carriers was introduced.

The key breakthrough in unification came in 1967 when Steven Weinberg, then of M.I.T., and Abdus Salam, a Pakistani physicist working in London, independently applied a concept developed earlier by Sheldon Glashow to the weak force and electromagnetic interaction. The theory essentially argued that the weak and electromagnetic interactions are really the same thing, but appear to differ because the force-carrying particle has a mass in the weak interaction but is massless in the electromagnetic interaction. The force-carrying particle is always a boson. For the electromagnetic interaction, the force-carrying particle is the photon. The photon has zero mass, and thus can travel for any distance at the speed of light without degrading. The weak interaction, however, is carried by a massive boson, which cannot reach large distances before degrading.

It had been noted somewhat earlier by Feynman and Gell-Mann among others, following the ideas of Enrico Fermi, that a boson of sufficiently large mass could explain the short range of the weak interaction. Feynman and Gell-Mann used only charged bosons in their theory; so they thought the weak interaction could occur only when one particle changed its charge by one unit and emitted the charged weak boson, and another particle absorbed the weak boson and also changed its charge. In other words, they thought that the weak interaction required a change in electric charge. This is called a charged current. The photon, having no electric charge, does not change the charge of the emitter or absorber, and is called a neutral current.

Weinberg and Salam's theory said that the only difference between the weak and electromagnetic interactions is the mass of the force-carrying boson—zero for electromagnetic, very heavy (about a hundred times that of a proton) for the weak interaction. In effect, they said that the weak boson was just a heavy photon. However, guided by the fact that the electromagnetic force carrier is neutral they suggested that there should be *both* neutral-current *and* charged-current weak interactions. The force-carrying charged bosons were labeled W^+ and W^-, and the neutral one was labeled Z^0. They were called as a group the *intermediate vector bosons*.

The major testable predictions of the theory were the existence of neutral currents and the existence of massive bosons. In a series of experiments carried out by various experimental groups at CERN and at Fermilab, it was established by 1973 that the weak interaction sometimes occurred without any change of electric charge. Thus, neutral currents existed. The Weinberg, Salam, and Glashow theory was also able to predict the masses of the W and Z particles. Although these predictions were important for physics, it was the elegance of their unification ideas and the subtlety of the experiments that verified them for which Weinberg, Salam, and Glashow were awarded the Nobel Prize in 1979.

Unification of the weak and electromagnetic forces put the force carriers W^+, W^-, Z^0, and the photon into the same family. The fact that three carriers of the family are heavy and one has zero mass accounts for the apparently different behavior of the forces. This has raised a very deep question: Where does mass arise? What determines mass?

THE CERN COLLIDER, THE FERMILAB TEVATRON, AND THE ELECTROWEAK BOSONS

Before the existence of the W and Z bosons could be proved, higher-energy accelerators were needed. In the late 1970s the two large proton accelerators at CERN and at Fermilab launched ambitious and divergent improvement programs. At CERN the project was to convert the 400-GeV accelerator into a storage ring that would accelerate, store, and collide counterrotating beams of protons and antiprotons. The drive was led by Carlo Rubbia, who was obsessed with the notion of "finding the W." This hypothetical massive particle was a key prediction of the electroweak theory. Because it was expected to be about 100 GeV in mass, it could not be produced by the 400-GeV beams because the laws of conservation of energy and momentum limited the energy available for producing new particles to only 27 GeV.

Rubbia reminded the CERN management that head-on collisions provided much more energy. If the CERN machine could accelerate protons to 400 GeV, antiprotons coming at them from the opposite direction would generate collisions producing 800 GeV—more than enough to make W's. The technical problems were formidable. Rubbia had to create a source of the antiprotons (just recently considered exotic) in unprecedented numbers, inject them into the accelerator complex, and

The two concentric rings of the Fermilab antiproton source. The outer (debuncher) ring receives 8-GeV antiprotons from a target bombarded by 120-GeV protons. After radio-frequency treatment to reduce the energy spread and to decrease the transverse motions, the antiprotons are transferred to the inner accumulator ring, where they are "cooled" and accumulated at a rate of 2×10^{10} per hour. A stack of 10^{11} antiprotons is sufficient for transfer back to the accelerator.

carefully accelerate them, simultaneously with protons, to the highest energy the magnets could be held to. The water-cooled magnets were designed to stay at a high field strength for only one second; now they would have to stay at that strength for the many hours during which collisions would take place. It was found that the highest field that could be successfully cooled would only enable CERN to reach 260 GeV, not its full 400 GeV. But 520 GeV (the collision energy of such particles colliding head on) was still enough to produce W's.

Meanwhile, at Fermilab, the goal was to raise the energy of the 400-GeV machine to close to 1,000 GeV. To do this, a ring of 1,000 superconducting magnets, 6 kilometers in circumference, would be added under the original machine. A beam of particles would be transferred from the old machine at 150 GeV to the superconducting ring, where it would be accelerated to 1,000 GeV. The laboratory could then extract this beam to use in fixed-target research or convert to a storage-ring collider, using protons and antiprotons like at CERN. The super-magnets do not get hot, and can be left for days at peak field of 1,000 GeV, so that 2,000 GeV would be available for exploring "new physics." This program was begun in July 1979.

Both programs worked. Both programs introduced fundamental improvements in accelerator technology. CERN pioneered the antipro-

An aerial view of the Fermi National Accelerator Laboratory. The largest circle is the main accelerator, with a circumference of 6.28 kilometers. Three experimental lines extend at a tangent from the accelerator. The small circle in front of the tall building is the 8-GeV Booster accelerator, which injects protons into the large circle.

ton source, made possible by a process called *stochastic cooling* invented by Simon Van der Meer of CERN. The trick was to produce antiprotons by collisions of protons extracted from the injector ring, at CERN, the 25-GeV Proton Source (PS). The antiprotons would be transported to a special storage ring where they would be accumulated over many hours. Because the "room" is limited in the vacuum chamber having a high-quality magnetic field, the antiproton beams must be compressed so that as many antiprotons as possible will fit. Stochastic cooling was a complex scheme for making the antiprotons move in a much more coherent and orderly manner, by the use of radio-frequency fields. In this way many more antiprotons could be added to the storage ring.

Fermilab pioneered the technology of using superconducting magnets for accelerators. Developed in the 1960s and 1970s, superconducting magnets make use of alloys of niobium cooled to about 4 degrees Kelvin (4 degrees above absolute zero). These magnets more than doubled the energy of the accelerator, while greatly reducing the use of electrical power. The Fermilab machine, known as the Tevatron, would become a prototype for a new generation of accelerator, the Superconducting Supercollider (to be discussed in Chapter 7).

The final proof of electroweak unification took place in 1983, when Carlo Rubbia, Simon Van der Meer, and their team of 130 physicists

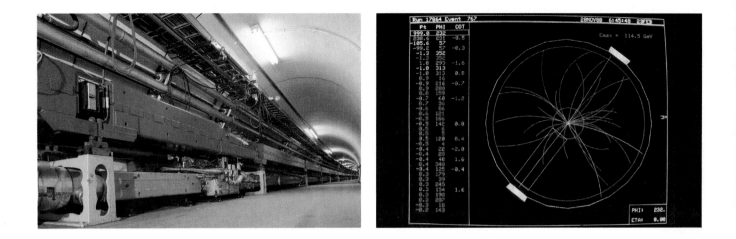

Left: The tunnel enclosing the Fermilab Tevatron. The upper string of magnets (red and blue) is the original 200-GeV accelerator completed in 1972. The lower, yellow string are superconducting magnets, which can accelerate protons to almost 1000 GeV. *Right:* A Fermilab CDF computer reconstruction of a Z^0 boson decaying into an electron and a positron back-to-back. The electron and positron are the fairly straight (because high momentum) yellow tracks "pointing" to the pink rectangles, which indicate a large energy release in the electromagnetic calorimeter. The energy of the electron and positron add up to the mass of the Z^0 particle. The other, lower-energy tracks are "spectators" to the main event.

actually produced the intermediate vector bosons, including the neutral Z^0. In fact, the experimenters found precisely the masses predicted by the electroweak theory. In 1985, Rubbia and Van der Meer shared the Nobel Prize for their discovery of the W and Z particles. The table on the facing page summarizes our knowledge of the electroweak force and the other forces in the standard model.

SYMMETRY

Our understanding of how fundamental forces can combine at high energies is closely bound up with the concept of symmetry. Imagine a piece of the microworld in action: Particles smash into one another; matter is materialized out of radiant energy; particles absorb quanta; particles emit quanta; here a hyperon, produced in a violent collision, spontaneously explodes into a shower of debris; there antimatter and matter annihilate one another into a spray of such particles as pions and kaons. An observer sees what appears to be a chaotic frenzy of meaningless activity. A physicist, trained, ever hopeful, and biased to search for simplicity, strives to discern some order, some regularity in these processes. A metaphor for these processes could be a soccer game with the ball deleted. The apparently aimless running about of twenty players would mystify extraterrestrial observers, until someone more brilliant than the rest hypothesized the existence of a ball and an objective; then all sorts of rules and order would suddenly become apparent.

The regularities in the behavior of particles are called *symmetries,* and they are closely related to conservation laws. Symmetry is related to

THE FORCES IN THE STANDARD MODEL

Force	Range	Strength at 10^{-13} centimeter in comparison with strong force	Carrier	Mass at rest (GeV/c²)	Spin	Electric charge	Remarks
Gravity	Infinite	10^{-38}	Graviton	0	2	0	Conjectured
Electromagnetism	Infinite	10^{-2}	Photon γ	0	1	0	Observed directly
Weak	Less than 10^{-16} centimeter	10^{-13}	Intermediate bosons: W^+	81	1	+1	Observed directly
			W^-	81	1	−1	Observed directly
			Z^0	93	1	0	Observed directly
Strong	About 10^{-13} centimeter	1	Gluons g	0	1	0	Confined, observed indirectly

(a) A Greek temple illustrates translational symmetry. If the row of columns is shifted by exactly one column spacing, the new temple is unchanged. (b) A circle illustrates rotational symmetry through any angle. If the dot is invisible, then the circle is not changed by rotation through any angle. (c) A three-bladed propeller illustrates 120° symmetry. Only rotations through 120° produce an invariant configuration.

the concept of *invariance:* If a change made in a physical system produces no observable effect, the system is said to be invariant to the change, implying a symmetry. The entire idea is most easily visualized in geometric systems.

Consider a stately row of elegant Corinthian columns, 9 meters apart, stretched out along an endless line, as in the figure on this page. Shift the entire row by exactly one column spacing. No change! The obvious symmetry implies also that the system is invariant to the shift operation. Mathematically, we say that if we replace all values of x by $x + 9$ meters, the various equations that describe nature in this simple one-dimensional world will not change. Rotation of circles through any angle, rotation of perfect spheres through any angle and about any axis,

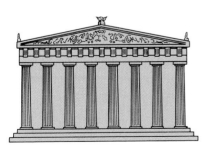

(a) Translational symmetry

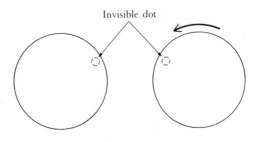

Invisible dot

(b) Reflection symmetry

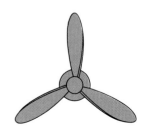

(c) Rotational symmetry

and rotation of squares by 90° are other examples of systems demonstrating symmetry and invariance to certain special operations.

The connection between conservation laws and invariance principles was made in the 1920s by an eminent mathematician, Emmy Noether (1882–1935), working in Göttingen, Germany. (The discoverer of the profound ideas ushering in a new view of the deep significance of conservation laws was not even allowed to lecture in the male-dominated society of that time!) She saw that the law of conservation of momentum implies invariance of all the laws of physics to *a spatial displacement*. The general Noether theorem asserts that if all values of the spatial coordinate x are replaced by $x + A$ (i.e., if the entire system under study is shifted by A units in the x direction), then the x component of the total momentum is conserved in all reactions. In a global extension to the physical universe, the conservation of momentum implies a deep symmetry or simplicity: There is no preferred location; space is smooth, homogeneous. The laws of physics, the results of all possible experiments, do not depend on the absolute location in space of the laboratory. Thus, all the equations describing all the events must be so structured that when x (as well as y and z) is changed by adding A, the value A simply cancels everywhere.

In a similar way Noether showed that the conservation of angular momentum (a measure of the rotation activity in an object) is related to the symmetry of direction in space, *isotropy*. All directions are the same, and if every direction in a physical system is rotated by, say, B degrees, the laws of physics do not change; they are invariant, and B drops out of the equation. Likewise, conservation of energy implies a symmetry in the flow of time and invariance of the equations (and the laws of physics) to a change in absolute time. As in the continuous operations of rotation and translation, absolute location, absolute position, and absolute time are not relevant if the experimentally observed conservation laws are indeed universally valid.

Nonspatial conservation laws are known and teach us about new symmetries. For example, the law of conservation of electric charge, long known from experiment, was verified again in the many particle reactions in which the total charge is always exactly constant. The corresponding symmetry applied to the equations of electromagnetism is called *gauge symmetry*. It insists that a correct mathematical description of electromagnetism must be invariant to certain transformations of quantum-mechanical quantities. When we have imposed this symmetry, the theory then correctly predicts conservation of charge. But an additional depth appears: All of the equations of electromagnetic theory, that is, Maxwell's equations subject to quantum mechanics (quantum electrodynamics, QED), can be derived from the gauge symmetry! So deep is the symmetry idea that we now believe that all the other forces in nature obey a form of gauge symmetry.

After the discovery of the quantum theory, new symmetries appeared that related to the quantum properties of atomic-scale systems. Here, it was soon discovered that symmetries could be imperfect: They could be "broken." Historically, the most spectacular example had to do with the three discrete symmetries parity (P), time reversal (T), and charge conjugation (C).

Parity symmetry is the notion that nature on the atomic level does not distinguish between left- and right-handedness. In the macroscopic world we have right-hand screws (rotation clockwise produces forward motion) and left-handed people. Parity symmetry is sometimes called mirror symmetry (in a mirror, right and left are interchanged), and the statement could be rewritten as: Nature cannot distinguish between the world and the mirror world. For example, suppose we filmed a series of experiments in a laboratory, one wall of which is a mirror. The test of mirror symmetry is whether we could tell if the camera was recording the experiments in the laboratory or in the mirror image of the laboratory. The laws of physics were believed to be invariant to the transformation $x \rightarrow -x$ and $y \rightarrow -y$ if z is the coordinate perpendicular to the mirror. This transformation is symbolized by parity symmetry P.

In 1956, the long-standing validity of the P symmetry was challenged by two young Chinese-American professors, T. D. Lee and C. N. Yang. In 1957, experiments carried out on the decay of pions, muons, and radioactive cobalt 60 proved that P symmetry was not respected by the weak force (see the figure on the next page).

Charge conjugation (C) symmetry exists between particle and antiparticle. Until 1957, it was thought that no experiment could distinguish a system composed of particles from a system composed of antiparticles. Again, the 1957 experiments demonstrated that C symmetry is also violated by the weak force.

After the fall of the P and C symmetries, it was noted that the *combined* operation of CP was in fact respected by the same experimentally observed processes that had separately broken C and P. The results of the pion and muon experiments seemed to confirm the combined CP symmetry until the 1964 experiments carried out at the Brookhaven AGS by Val Fitch and James Cronin, then at Princeton, on the decay of neutral kaons. Here clear evidence of the failure of CP symmetry was obtained: The spontaneous decay of neutral kaons into pairs of pions supplied crucial evidence that the weak force does not respect CP symmetry. The mirror process involving antimatter turned out to be quite different from the laboratory process.

Finally, there arose a question about the *direction* of time flow. Let's imagine an experiment (it is easier to think of a simple one like the collision of two particles, but it could also be much more complicated) in which we reverse the sign of t (time) in all equations describing the experiment. If, for example, t only appears as t^2, then setting $t \rightarrow -t$

Parity violation is a violation of mirror symmetry. (a) To preserve mirror symmetry, a spinning muon, upon decay, would have to emit equal numbers of electrons in both the upward and downward directions. (b) However, the real muon emits more electrons in the direction of the spin, if the muon is thought of as a right-hand screw, advancing as it rotates. Thus, muon decay violates mirror symmetry, since a true mirror image would emit electrons *against* the spin direction. The "mirror image" shown is nature's muon, that is, the original simply turned upside down. A true mirror image muon does not exist, illustrating that the weak force does not respect parity or mirror symmetry.

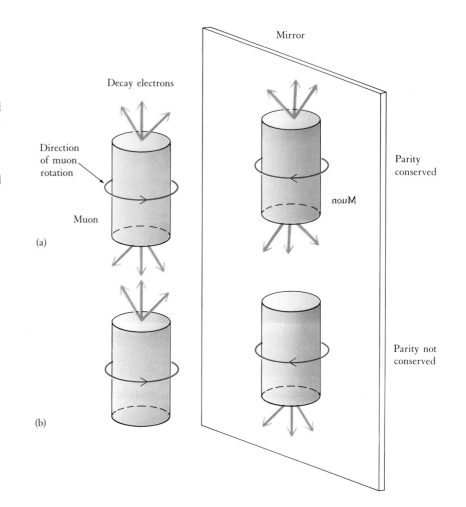

produces no change in the equations, no change in the course of the experiment. A movie of the experiment could be run backward or forward; both would give possible processes. Like many quantum ideas, this is counterintuitive, but the invariance with respect to the time reversal (T) operation ($t \rightarrow -t$) seemed iron-clad until the Fitch–Cronin experiment.

Years before this, Pauli had pointed out that a sequence of operations like CPT could be imagined and studied; that is, in sequence, change all particles to antiparticles, reflect the system in a mirror, and change the sign of time. Pauli's theorem was that all nature respected the CPT operation, and in fact that this was closely connected to the relativistic invariance of Einstein's equations. There is a consensus that CPT invariance cannot be broken (at least not at energy scales below

10^{19} GeV). If any young experimenter ever succeeds in demonstrating the failure of CPT symmetry, he will have the sanity of thousands of physicists on his conscience! However, if CPT is a valid symmetry, then when Fitch and Cronin showed that CP is a broken symmetry, they also showed that T must similarly be broken.

Some of the quantum conservation laws that are related to important symmetries are conservation of charge, of lepton number, and of baryon number. As an example of the latter, until the 1970s it was thought that the total number of baryons (protons, neutrons, lambdas . . . i.e., all hadrons made of three quarks) in the universe is conserved. More precisely, the number of baryons minus the number of antibaryons is a constant. This implies that protons live forever, but that a proton plus an antiproton can be annihilated or produced by an energetic collision. Massive experiments built underground were designed to detect proton decay but failed. The resulting proton lifetime was longer than 10^{32} years! The instability of the proton is a theoretical conjecture supported by cosmological arguments we will discuss later in the book. Clearly, proton nonconservation would imply quark nonconservation, meaning that quarks would be able to change to leptons.

SYMMETRY BREAKING: THE HIGGS PARTICLE

Electroweak unification is based on the concept that forces are generated by (and manifested as) the quantum exchange of bosons. That is, to say that there is a force between two fermions is to say that they are exchanging bosons. The key to modern unification ideas is in understanding how some of the fundamental force-carrying bosons acquire mass. For electroweak unification, the question is how did W^-, W^+, and Z^0 acquire masses of about a hundred times that of a proton, whereas the photon remains massless? In physics terms, we say that the symmetry between the weak and the electromagnetic interaction is *broken* by the mass of the intermediate-vector bosons.

We noted earlier that gauge symmetry first appeared as the symmetry related to the conservation of electric charge in electromagnetic interactions. Gauge symmetry proved to be more profound than the theory that spawned it and turned out to be a crucial key for understanding all interactions in the universe. One curiosity about gauge symmetry is that it requires the force carriers (now called gauge bosons) to be massless. This requirement was satisfied by the photon but not by the W and Z heavyweights of the weak force, which also had to obey gauge symmetry. To get around this obstacle, Salam, Glashow and Weinberg had to invent a mechanism for breaking the symmetry—that is, for allowing the W and Z, considered really massless, to acquire mass.

Such a mechanism had been developed in other contexts by various theoretical physicists, including Yoichiro Nambu of Chicago, Jeffrey Goldstone of M.I.T., Glashow, the Scottish physicist Peter Higgs, and Phillip Anderson of Princeton, among others. The idea is to postulate the existence of a new field, which became known as the Higgs field. Unlike the gauge bosons, this field has no direction or spin associated with it and therefore belongs to a class of (yet to be experimentally discovered) fields called scalar fields. The Higgs field is different from other fields: At low temperatures space would rather be filled with Higgs particles than not, owing to the way Higgs particles interact with themselves. Today we are all living within a sea of Higgs particles. Because the W and Z (but not the photon) interact with the Higgs, they move through Higgs filled space as if they are moving through molasses. When moving through the Higgs molasses, the W and Z no longer move at the speed of light, and they can be described as having acquired an effective mass. At high temperatures and energies, the Higgs interactions are such that space is no longer filled with Higgs molasses and the W and Z no longer have to slough along. Thus at high energies the W and Z lose their mass, and the symmetry between W and Z and the photon becomes manifest. (When we discuss the early universe in Chapter 6, we will find that as the universe approaches the critical transition from high to low temperature, the universe finds itself in a "false vacuum" where space itself appears to have a net energy. This false vacuum is due to the Higgs field.) In the case of the now unified electromagnetic and weak force, low and high energy are relative to about 100 GeV, the scale of the electroweak force. Sounds bizarre? There is more! The Higgs field as so invented also gave masses to the quarks and leptons, thus addressing one of the deepest problems in particle physics: the nature of mass. Why is the muon, so similar to the electron in all other properties, so much heavier? Already posed back in the 1930s, the questions gave rise to exhaustive experiments designed to find some structural clue to the mass difference between electrons and muons, all to no avail. Then, with the discovery of quarks, the puzzle of mass only became worse. Although the Higgs phenomenon certainly did not solve this problem, it created an encouraging mechanism by means of which this 50-year-old puzzle could be restated. Particles acquire a mass because of the Higgs field, but why each particle acquires a different mass or, like the photon, acquires no mass, was left open. The question could now be rephrased in terms of the strengths of the affinity of different particles for the Higgs field.

The Higgs field was postulated in electroweak theory in order to preserve the symmetry at high energy and to hide or break the symmetry at low energy. We may perhaps clarify this by an analogy. Consider an array of tiny molecular magnets at high temperature. They are moving rapidly by virtue of their high temperature and each of the magnets points in an arbitrary direction. Symmetry in space is complete, since no direction is preferred. Lower the temperature, and sooner or later, a subgroup will align in *one* arbitrary direction and will force all the others to do the same. Now we have a *magnet,* and symmetry is destroyed. It can be restored by raising the temperature (energy). The transition from symmetry to broken symmetry is caused by the lowering of the energy.

Obviously, a critical experiment is to get the energies high enough that we can produce a physically observable Higgs field or the quantum equivalent—the Higgs particle. No experiment has done that yet. Finding a Higgs particle would verify our concepts of unification and allow them to be extended to the higher-energy realm where the strong interaction and gravity might also become involved. Since the Higgs field fills space even when no particles are present, we might think of it as being like the ether that nineteenth-century physicists thought was necessary for electromagnetic wave propagation. And like the ether, if the Higgs field is *not* discovered, that non-discovery will also revolutionize physics. As we will see, the cosmologists were forced to invent a field very much like the Higgs field in order to solidify the theory of the big bang. For these reasons, the spotlight will be on the Higgs concept in the 1990s.

SUMMARY

In this chapter we have seen the discovery of the standard model of elementary particles now used to describe the world of inner space. This model is summarized in the table. We have seen how experiments, which reached such energies that they were probing distances smaller than a single proton, found that the protons and neutrons are not the fundamental, indivisible building blocks of nature they were originally thought to be. Instead, protons, neutrons, and the mesons that were thought to hold them together are all made out of some much smaller, pointlike particles known as quarks. These quarks come in various flavors that signify their symmetry properties, and they come with three strong charges, known as colors. The force between the quarks is such that quarks can never get out of subatomic particles except perhaps at the birth of the universe (or in the collapse to black holes, where they would remain trapped).

Experimenters developed electron–positron colliding machines that found an entirely new quark, charm, in the form of a novel atom composed of a charmed quark bound to a charmed antiquark. This was the J/ψ. Theorists had speculated on the possibility that another quark

The standard model of particle physics contains six colored quarks and six colorless leptons, which are subject to the forces of nature carried by twelve experimentally established gauge bosons and one hypothesized graviton. The gluons each carry a color—for example, blue—and an anticolor—for example, antired (shown here as red). Although there would apparently be three gluons shown as white because of the cancellation of, say, red and antired, one of these is mathematically superfluous, and so we have eight gluons instead of nine.

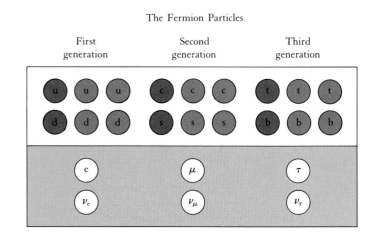

The Fermion Particles

First generation   Second generation   Third generation

Quarks

Leptons

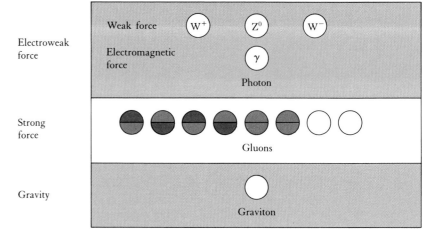

The Boson Force Carriers

Electroweak force

Weak force — W⁺ Z⁰ W⁻

Electromagnetic force — γ Photon

Strong force — Gluons

Gravity — Graviton

flavor existed. The discovery of charm provided additional input into quark dynamics that enabled the strong interaction between quarks to be figured out, and provided an explanation for the absence of free quarks. The 1976 discovery of a third charged lepton, τ, and of the upsilon state of beauty, or bottom, quark and antibeauty quark put to rest any doubts and firmly established the third generation of particles. These discoveries came out of new accelerators at SLAC and Fermilab.

The standard model asserts that all matter is made of twelve fermions subdivided into six flavors of quarks and six flavors of leptons. These particles of matter interact with one another via four forces. Excluding gravity, we have identified twelve force-carrying bosons: one photon; the W^+, W^-, Z^0 particles; and eight gluons. These bosons carry the electromagnetic, weak, and strong forces, respectively.

We've also seen in this chapter how the fundamental forces of nature are beginning to be unified. In particular, experiments have now verified that the weak and electromagnetic forces are really one and the same at high energies. We call this new unified force the electroweak force. A price for this unification was the postulated existence of new particles, called Higgs particles, creators of a field that permeates all space and influences the properties of the vacuum. The success of this unification has enabled physicists to conceive of more complete unification theories, such as "grand unified theories" (GUTs), which would unify the strong interaction with the electroweak, or even "theories of everything" (TOE) which would include gravity as well. We will return to these ideas and their possible tests in later chapters.

In all of this, we have seen an interplay of experiment and theory. Sometimes theory leads the way and makes predictions; at other times experiment discovers something totally unexpected. Frequently an experiment finds a vague theoretically predicted object, but gives it substance with specific mass and properties. Science seems to progress best when theory and experiment interact closely. If theoretical predictions are in untestable regimes or if some mass of data is unclassifiable, then science moves slowly. But when each stimulates the other, rapid progress occurs, as it did during the 1960s and 1970s, when the quark model developed.

Today, this close interplay of experiment and theory is encompassing the subdiscipline of cosmology and outer space as well as inner space. As we will see in the next chapter, problems such as determining the number of fundamental particles are being addressed by cosmological theory as well as particle physics. These answers, in turn, can be checked with accelerator experiments rather than just the observations that have traditionally characterized astronomy. Furthermore, as accelerators go to higher energies, they are in effect duplicating the conditions of the hot early universe. Thus, accelerators give us clues about how the early universe behaved, which in turn might illuminate problems in cosmology.

5 THE ESTABLISHMENT OF THE BIG-BANG THEORY

As the last few chapters have described, our understanding of the nature of matter, or inner space, has changed greatly during most of the twentieth century. Paralleling these changes were tremendous developments in our knowledge of the universe as a whole: outer space. Until the 1970s, these two areas, inner space and outer space, were considered quite separate. Astronomers used telescopes and looked at big things; nuclear and particle physicists used accelerators (which served as microscopes) to look at small things. However, with the establishment of the big-bang theory (proof that the universe was at one time extremely hot and dense), it became clear that studying the very small affected our understanding of the very large, and conversely, that studying the very large gave us important information about the very small. Let us, therefore, survey the twentieth-century developments in cosmology.

TELESCOPES

The beginning of the twentieth century saw a great advance in telescope design and construction. The largest refracting (lens) telescopes had been built at the end of the nineteenth century: the 40-inch telescope at Yerkes Observatory in Williams Bay, Wisconsin, operated by the University of Chicago, and the 36-inch telescope at Lick Observatory, in California, operated by the University of California. Refracting telescopes use a lens to create an enlarged image, whereas reflecting telescopes use a curved mirror. Although Isaac Newton had built the first

Astronomers can study objects in space by detecting other forms of radiation besides light, as demonstrated by this radio telescope map of the spiral galaxy M83. The color denotes the density of atomic hydrogen.

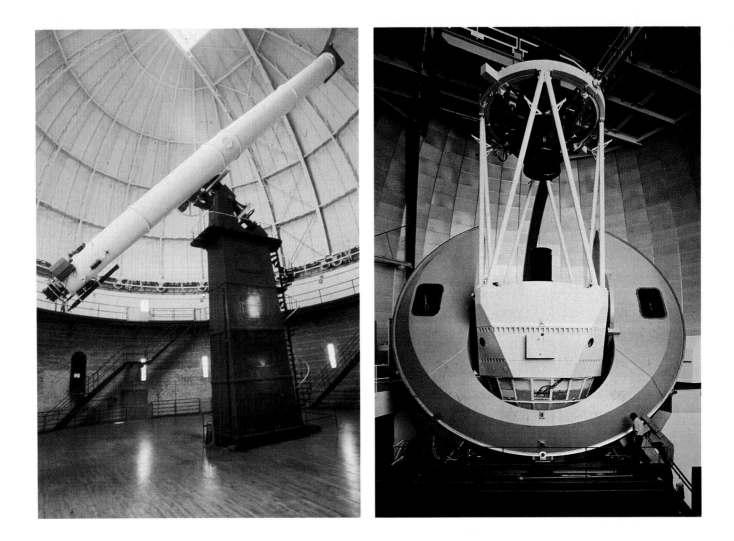

Left: The world's largest refracting telescope, located at the University of Chicago's Yerkes Observatory in Williams Bay, Wisconsin. The diameter of the primary lens is 40 inches. *Right*: This 4-meter-diameter reflector is the largest optical telescope at the Kitt Peak National Observatory near Tuscon, Arizona.

reflecting telescope in 1672, refracting telescopes remained more popular until their technical limits were reached at the end of the nineteenth century. The ensuing shift to reflectors enabled a major improvement in telescopes' light-gathering capability. The light-gathering capability is directly related to the surface area of the lens or mirror, which is proportional to the square of the diameter: $(200/40)^2 = 25$. As the picture of the Yerkes refractor shows, a refracting telescope is a very long tube that requires substantial structural support, even if it only has 40 inches of light-gathering capacity, whereas the Palomar 200-inch reflecting telescope, though not nearly as long, has twenty-five times more light-gath-

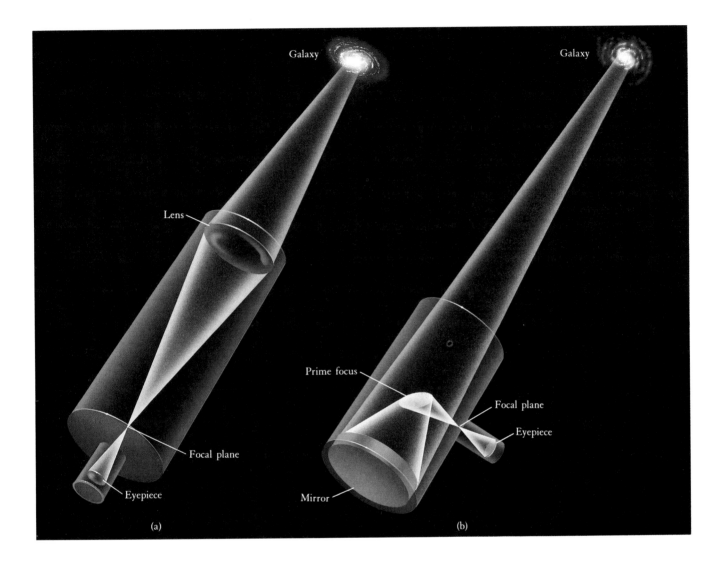

(a) A refracting telescope uses two lenses. The larger, convex lens focuses the incoming light rays at the focal plane, where an image of the astronomical object is formed. The smaller lens serves as the eyepiece and is used to magnify and examine the image. (b) A reflecting telescope depends on a curved mirror to focus the light rays. A smaller mirror (the prime focus) deflects the light rays to one side where an eyepiece magnifies the image.

ering capability. Making a lens 200 inches in diameter was not technologically feasible, but a mirror of that size was quite reasonable. The art of mirror making is by no means trivial. Very special glass must be formed in a casting which, for a large solid mirror, takes many months to cool safely. The polishing process is designed to form a surface to a precision approaching the wavelength of light. This takes years. Finally, the entire mechanical system must be able to rotate the massive mirror and then hold it steady for long exposures with exquisite precision.

The telescope that opened up twentieth-century cosmology was the 100-inch reflector constructed by George Ellery Hale (1868–1938) on Mt.

A charge-couple device. It occupies an area of 60 square millimeters.

Wilson near Pasadena, California. Hale had helped develop Yerkes Observatory and was later able to raise funds from the Carnegie Foundation to build a telescope on a mountain in the West, where weather would not be as much of a problem as it was in the Midwest.

In addition to needing large telescopes to gather the light from very faint objects, astronomers needed a more sensitive device for detecting that light as it came through the telescope. The original telescope detector was the human eye. An observer would look through the telescope and sketch what he saw. Obviously, this technique was flawed by the limits of human memory and by the desire of the eye to find patterns where there were none (for example, the purported discovery by the astronomer Percival Lowell of canals on Mars). However, by the end of the nineteenth century, a detector came into existence that was less biased and less limited than the human eye, namely, photographic film.

Throughout the twentieth century, photographs have been one of the main ways of cataloging the heavens. Not only is a photograph able to tell something about the position of an object, but it also tells something about the intensity of the light coming from the object. Intense light will chemically burn out a larger area around an object's position on a photographic plate. In this way photographs also served as photometers—devices for measuring the intensity of light. However, photographic photometers were eventually replaced by electronic photometers, in which the photons trigger electronic signals that can be directly counted as opposed to the less precise method of merely causing a chemical reaction in the photographic plate. Modern telescope detectors have developed electronic photometers even further, with the use of solid-state electronics known as charge-coupled devices, or CCDs. These devices can directly measure the number of photons hitting a given part of the focal plane of a telescope. Thus, in some sense, they serve as electronic photographic negatives, but with far more sensitivity to weak light than the chemical mix on a photographic plate.

In addition to the intensity of light, astronomers also need to know the spectrum of light emitted by an object in order to see an object's color and whether atomic lines are present or absent. Photographic detectors were able to do this by putting a prism in front of the photograph, thus spreading the light out into its separate colors. Spectra were thus obtained for stars almost as soon as spectra were obtained for chemical elements in the laboratory. It was discovered by these techniques that the lines found in the spectra of elements in the laboratory and those of the sun were also found in the light coming from stars, thus verifying that the same chemical elements exist in the stars. Again, with the development of electronics, electronic detectors replaced photographs on spectroscopes, but the basic technique remained the same.

As Hale completed the 100-inch telescope at Mt. Wilson, he was able to persuade a former Chicago student, Edwin Hubble, to join him

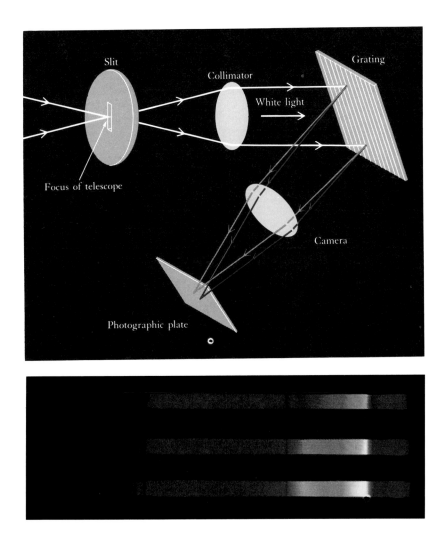

Top: This type of spectroscope uses a device called a diffraction grating to break starlight into separate colors. The grating is a piece of glass onto which thousands of closely spaced lines are cut. Light rays reflecting off the various parts of the grating interfere with each other so as to produce a spectrum. The collimator is a lens that ensures that light rays hitting the grating are parallel. *Bottom*: These spectra show light from the star Vega broken into its constituent colors.

in observing on this California mountaintop. Although Hale was the entrepreneur who was able to construct the great instruments that dominated world astronomy for most of the twentieth century, it was Hubble (1889–1953) who used the instruments to make the critical observations that told us about the universe. Hubble did not start out as an astronomer. In college his athletic prowess was legendary. He studied law at Oxford as a Rhodes scholar and, after some teaching and law practice in Louisville, he "chucked" it all and returned to the University of Chicago to study astronomy.

As World War I ended, Hubble was making observations on the nature of the spiral nebulae. A major debate at that time was whether or

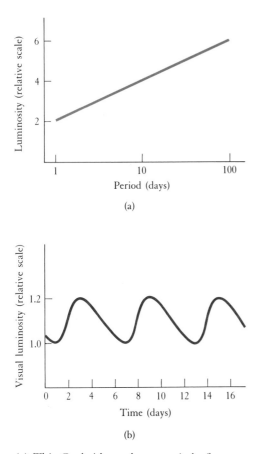

(a) This Cepheid star has a period of variation of about six days. (b) The average luminosity of classical Cepheids can be predicted if their period of variation is known.

not these fuzzy patches in the sky were clouds of gas in our Galaxy or some much more distant object. Hubble established that the spiral nebulae were indeed external galaxies, many of them the size of our whole Milky Way Galaxy. Hubble took his first step toward establishing the true nature of spiral nebulae when he noted that they contained little dots of light blinking on and off. Some years before, Henrietta Levitt, working at the Harvard Observatory, had shown that there was a certain class of star, known as a Cepheid, whose light output varied. Levitt established that the period of variation was directly related to the intrinsic luminosity of the Cepheid. Hubble found that these dots of light in the nebulae were varying with the period of Levitt's Cepheids, but appeared to be much dimmer. He concluded that they must indeed be the Cepheids of Levitt, but were at such enormous distances that their luminosity appeared very low. Then, using the relationship that the intensity of an object's light is inversely proportional to the square of its distance from the observer, he was able to calculate the distances and show that they were far beyond any other star in our Galaxy and thus must be in external galaxies. In particular, he established the distance to the Andromeda nebula. Andromeda is the nearest major galaxy to us, and is now known to be at a distance of 2 million light-years; whereas the radius of the disk of our Galaxy is now known to be only about 25,000 light-years. Prior to that time, it had been assumed that the stars of the Milky Way encompassed the entire universe. Thus, Hubble's observation established that the universe was much larger than anybody had ever dreamed.

Hubble's contribution to our knowledge of the universe did not stop there. He went on to explore these external galaxies in more detail and found that almost all of them were moving away from us. He did this by measuring the shift of the wavelength of the atomic lines in these galaxies. In 1842, the Austrian physicist Christian Doppler had shown that the wavelength of any wave is affected by the velocity of the source of the wave. If the source moves toward the observer, the wavelength is compressed to a smaller length. If the source is moving away from the observer, the wavelength is stretched. The amount of stretch or compression can be directly related to the velocity of the source.

The atomic spectra that had been measured in laboratories were found to have the same pattern in astronomical objects, but all shifted in wavelength. Hubble was able to calculate the velocity of a galaxy by observing the shift of the galaxy's atomic lines relative to lines emitted from a stationary source and measured in a laboratory. He noticed that with the exception of nearby galaxies, such as Andromeda, all the galaxies in the universe had their wavelength shifted toward longer values. (The Milky Way, the Andromeda galaxy, the Magellanic Clouds, and a few other galaxies make up what is known as the Local Group of galaxies, which are gravitating around each other. The cosmological expan-

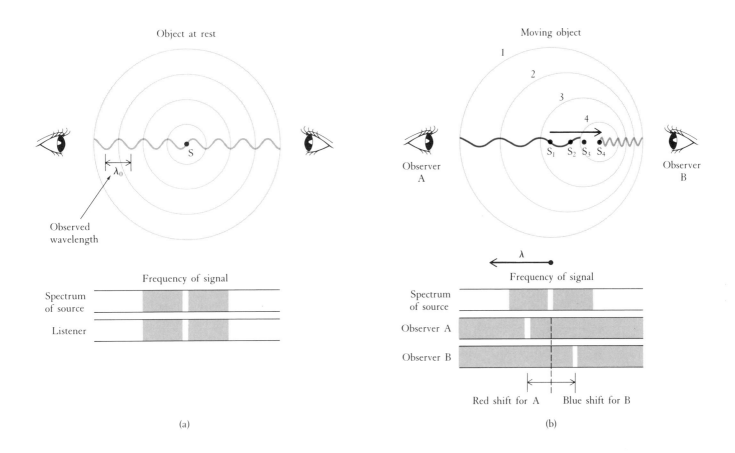

(a)

(b)

Observers see the Doppler shift when the velocity of an object in motion is small compared with the speed of the waves emitted by the object. (a) Observers on opposite sides of a stationary object will see waves of the same length λ_0. The waves move at the speed of light c. (b) An observer in the direction of motion will see a shorter wavelength, $\lambda = \lambda_0(1 - v/c)$, where v is the velocity of the object. An observer away from the direction of motion will see a longer wavelength, $\lambda = \lambda_0(1 + v/c)$. Circle 1 is the location of light emitted when the source was at S_1, Circle 2 is the location of light emitted when the source was at S_2, and so forth.

sion does not apply at the scale of the Local Group, just as it doesn't affect the distance between the Earth and the Sun, or apply to the structure of the Milky Way.) Because the longest wavelengths of optical light are red, and the shortest are blue, this shift of the wavelengths was referred to as a *redshift*.

Hubble was not the first to note that the spiral nebulae had atomic lines that appeared to be redshifted. In fact, V. M. Slipher, working at the Lowell Observatory, had noted as early as 1912 that many of the spiral nebulae showed this effect. However, until Hubble established that nebulae were at large distances from Earth, the interpretation was not obvious. To figure out systematically what was going on, Hubble needed redshifts not just on the galaxies that Slipher had done, but on many, many more, each of which had the distance determined. To get these large numbers of redshifts, he employed the help of L. Humason.

Humason was a colorful character. He was a mule skinner who trekked supplies 5,000 feet up the mountain from Pasadena to the Mt. Wilson Observatory. After each mule-train trip to the observatory, he would spend the night before going down. He was curious, and when walking around in the evenings while the mules were resting, he would inquire about what the astronomers were doing. He realized that he could do some of the measuring as well, and that it was interesting and fun to do. He started working with Hubble measuring redshifts of galaxies.

Hubble used these redshifts in his distance determinations to establish that the universe is expanding, that is, most of the objects were moving away from each other, as indicated by the fact that they had redshifts rather than blueshifts. He also established that the velocity v of expansion was proportional to the distance,

$$v = Hr$$

where H is a constant we now call the Hubble constant, and r is the distance of the galaxy from us. The farther away a galaxy is, the faster it is moving. This kind of expansion is very special because it enables shapes to remain the same. Most expansions do not obey Hubble's law. For example, when a firecracker explodes, it no longer looks like a firecracker, but like little shreds of paper. However, the Hubble expansion is very similar to a rising loaf of raisin bread, which as it expands continues to look like a loaf. The raisins move apart from each other, obeying Hubble's law. This can be understood with a loaf of raisin bread, because the raisins have more dough separating them. As the dough rises, the relative velocity of the raisins is greater for the raisins that are farther apart. In a similar way, more distant galaxies have more

The raisin bread model of the universe. As the loaf rises, it continues to look like a loaf and the raisins keep the same orientation although all raisins continue to increase their separation from one another. In a similar way, the universe retains its same basic configuration throughout its expansion.

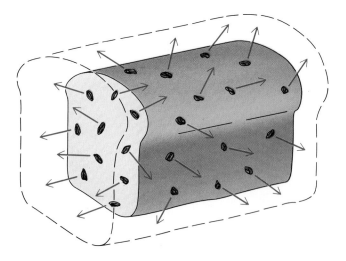

space separating them, and it is this amount of space that is increasing. Thus, the universe retains its same basic configuration throughout its expansion. It is important to stress that in the expanding universe, it is space itself with everything in it that is expanding.

The inverse of the Hubble constant gives us the time that it takes for a galaxy moving at its present velocity v to travel the distance r. Hubble first calculated this relationship in 1929. He found that this *Hubble time* $1/H$ was only 2 billion years. Since the Hubble time is a measure of the age of the universe, that is, how long it took to spread things out to the current state, it was very curious that his initial value for the time was less than the accepted age for the Earth. This led to a questioning of Hubble's interpretation of the Hubble constant and the expansion.

Resolution of this problem did not occur until the 1950s and 1960s, when it was realized that the scales used by Hubble were grossly underestimating the distances to the galaxies. Although it is relatively easy to get an accurate spectrum of a galaxy, thus determining redshift and velocity, distance is more difficult. We can measure a galaxy's light intensity. But to get a distance from that, we have to know what its intrinsic luminosity is, so that we can compare the intrinsic luminosity with how bright it *appears* to be, using $1/r^2$ to get a distance.

Unfortunately, finding out the intrinsic luminosity of distant astronomical objects is very difficult. The techniques that Hubble was using in 1929 had some very definite systematic errors. In particular, Walter Baade, also working at the Mt. Wilson Observatory, established in 1952 that Hubble's distance scale was in error by a factor of 2. Subsequent work by Hubble's associate Allan Sandage showed a distance scale that is five to ten times Hubble's distance scale. Sandage was able to do this by using the then new 200-inch telescope, built in 1948 on Mt. Palomar by the Carnegie Institute in collaboration with Cal. Tech. The 200-inch telescope, with its enhanced light-gathering power, was able to get spectra and details about galaxies much faster than Mt. Wilson's 100-inch telescope. Bigger telescopes have more light-gathering power. Therefore, an observer does not have to look at faint objects as long in order to collect enough photons to establish a spectrum and thus be able to determine a redshift. Furthermore, bigger telescopes can pick up fainter, more distant galaxies. In addition, the 100-inch Mt. Wilson telescope was already suffering from the light pollution of the Los Angeles basin, decreasing its usefulness. Today, the 200-inch Mt. Palomar telescope is beginning to have similar problems with light pollution from San Diego. (The best observations ever done at Mt. Wilson were made during the blackouts of World War II.)

The revisions of Hubble's distance scale yield values for the "Hubble-age" of the universe that are now between 10 and 25 billion years. Because of the continual uncertainties in the distance scale, we cannot yet reduce that range of uncertainty. It is hoped that when a high-quality telescope is placed into orbit around the Earth, more accurate distances will be obtained. In particular, the Cepheid stars might be seen in more distant objects, thereby establishing the distances to those objects. Such a telescope, appropriately named the Hubble Space Telescope, is scheduled to be launched in 1990. If the distances to more remote galaxies are better determined by using Cepheids, the uncertainty in the Hubble constant might be reduced considerably.

RADIO ASTRONOMY

Hubble's observations establishing that the universe is expanding were done using optical telescopes that use the same light that our eyes see. However, as we know from our discussions of electromagnetic interactions, electromagnetic waves can be at many different wavelengths and frequencies, not just the limited range to which the human eye is sensitive. We call longer wavelengths radio waves. Extremely short wavelengths are called X rays and gamma rays. Modern astronomy utilizes the entire range of electromagnetic radiation, from radio waves through microwave, infrared, optical, ultraviolet, X rays, and gamma rays, to look at objects in space. The development of nonoptical telescopes has been one of the modern revolutions in observational astronomy and has revealed much that optical astronomy was unable to show.

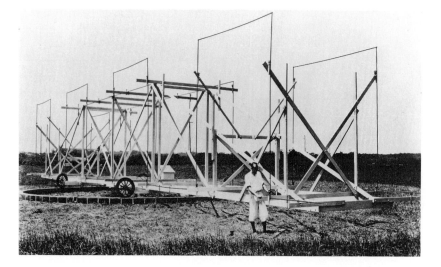

Karl Jansky standing next to the rotating antenna he used to discover radio waves coming from space.

Radio astronomy started in 1933, when Karl Jansky, working at the Bell Telephone Laboratories in New Jersey, pointed a radio antenna towards space. Jansky, a Bell Labs' radio engineer, had been asked to see if he could identify the origin of the static that interfered with radio communications. Jansky constructed a special antenna that could identify directions from which radio signals arrive. He adopted a wavelength of about 15 meters, then popular for communications work. Jansky's search for static led to a crucial discovery.

Underneath the prevalent noises from electrical equipment, thunderstorms, and so forth, was a steady hiss that persisted throughout his several years of measurements. Some intuition led Jansky to study books on astronomy, and by 1933 he had become convinced that the source of his hiss was extraterrestrial. The source seemed to be confined to the plane of the Milky Way. The origin of the hiss was naturally occurring radio emissions arriving at his antenna from stellar objects. Jansky's publication of the discovery of what Kepler had called "the music of the stars" attracted very little attention at the time, except for some press interest. *The New Yorker* magazine, commenting on a report of some engineer hearing a hiss from the Milky Way, said, "This is the longest distance anyone ever went looking for trouble." Jansky's desire to follow up on this was frustrated by Bell Labs' officials, who, once the source was known, lost interest in the annoying radio noise—a decision that must still be embarrassing to one of the preeminent research laboratories in the world!

In 1936, Jansky's discovery was seized upon by the self-taught inventor–astronomer Grote Reber. Working in his back yard in suburban Chicago, Reber showed that, indeed, there were very interesting radio waves coming from objects in space. In particular, Reber made a map of intense sources of radio waves, mostly from the Milky Way. Although the information startled many traditional optical astronomers, it was not accepted as anything of great significance until some unusual objects in space were found to emit tremendous amounts of radio waves. These objects, when looked at optically, were found to look like curious stars with unusual spectra, and were given the name quasi-stellar objects or *quasars*. Martin Schmidt, an optical astronomer working at the Palomar Observatory at Cal. Tech., noted that the funny spectral lines in the quasars could be understood if the objects were at extremely large redshift. This implied that they were at distances greater than most galaxies, if one uses the Hubble relationship that velocity is proportional to distance. Thus, the fact that quasars still looked like stars and yet were at these enormous distances implied that they were truly emitting a prodigious amount of radio and optical energy. Radio astronomy had therefore been used to find a whole new class of astronomical objects. It is now thought that these objects are associated with the birth of galaxies themselves.

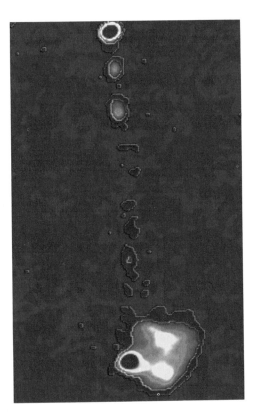

Radio telescope image of a quasar with a strong jet of matter streaming from it. The quasar's nucleus is located at the bright point near the top of the picture. Several knots can be seen leading to the lobe at the bottom of the picture. The colors are meant to depict the radio intensity and are not true optical colors.

STEADY STATE OR BIG BANG?

As was mentioned, the Hubble expansion was initially perplexing to some people because its timescale did not agree with other bounds on the age of the universe. In the 1940s, alternative models to simple, free expansion for the universe were proposed. In particular, a model known as the steady-state theory of cosmology was developed by Fred Hoyle, Herman Bondi, and Tommy Gold; in it new matter was created to fill in the space between the galaxies as the galaxies moved apart. This steady creation of matter would keep the density of the universe constant (in a steady state), but would also allow for the Hubble expansion.

The steady-state cosmology was important because it provided experimental tests for resolving whether the universe existed in a steady state or in the aftereffects of the big bang. According to the big-bang theory, the entire universe started out at very high density and temperature, and expanded and cooled to the present-day conditions of low average density and low temperature. The key difference between big-bang and steady-state cosmologies is that the density changes with time in the big-bang model, but remains constant in the steady-state theory.

By the late 1940s, George Gamow had noted that in its primordial hot and dense state, the big-bang universe would have produced radiation that should still be around, since nothing can escape from the universe. In the steady-state theory, however, the universe would never have had a high-density stage, and so would have no background radiation. (The term "hot big bang" was actually first invented by Fred Hoyle, who on a B.B.C. radio show used "big bang" as a term to describe Gamow's model, as contrasted to his own steady-state picture.) In 1965 Penzias and Wilson's discovery of the background radiation, which we will describe in the next section, helped establish the correctness of the big-bang theory.

Gamow and his colleagues Ralph Alpher and George Hermann also pursued the implications of the big-bang theory in terms of how the chemical elements would have been formed. He noted that the early universe would have been hot enough that nuclear reactions (nucleosynthesis) would have occurred everywhere, and that the products of these nuclear reactions might also still be around today. Originally Gamow thought that all the elements would be manufactured in the big bang. However, the facts of nuclear physics prevented this from occurring, since there are no stable nuclei with a total of either five or eight nucleons (neutrons and protons). In the big bang's sequential buildup from lighter nuclei, a nucleus with nine or more nucleons would have required a nucleus of eight nucleons to be made first. The details of nuclear reactions in the aftermath of the big bang could not easily bridge these instability gaps. As a result, big-bang conditions can make large amounts only of helium, as well as traces of other light nuclei. Stars, however, in contracting to high densities, have a way to jump over the gaps at five and eight nucleons, and so to synthesize all the heavy elements. This fact was viewed as an early triumph for the steady-state theory, because Hoyle advocated that all nucleosynthesis takes place in stars. Of course, stars also exist in a big-bang cosmology; so this theory was still in the running. It is now known that one quarter of the mass of the universe consists of a single nuclear species—helium 4. (Helium 4 is that isotope of helium having two neutrons and two protons, giving it a total of four nucleons. The other isotope of helium is helium 3, which has only one neutron with the two protons. Helium 3 is only produced in trace amounts in the big bang, but those trace amounts have also proven to be interesting.) There is no way in which stars could have produced this much helium 4. Helium abundance and the background radiation established that the universe did go through Gamow's hot big bang.

DISCOVERY OF THE COSMOLOGICAL BACKGROUND RADIATION

From a cosmological point of view, perhaps the most dramatic radio (and microwave) observation of all was the observation of the cosmological background. In the mid-1960s, two young postdoctoral researchers at the Bell Laboratories, Arno Penzias and Robert Wilson, were working with an antenna designed for microwave radiation (wavelength about 7 centimeters) to track Echo I, one of our very early artificial satellites. Other people had used this antenna to look for radio emission from objects in space. These previous scientists had merely adjusted the background of the antenna and looked for signals stronger than the background noise. Penzias and Wilson, on the other hand, were quite concerned that the background noise level was much higher than would have been expected for that antenna. No matter what direction the antenna was pointed in, the noise level remained the same. In an attempt to get rid of the noise, they removed a flock of nesting pigeons from the antenna and spent many hours scrubbing away the pigeon droppings, but still the noise persisted.

They discussed this mysterious noise with a radio astronomer, now at M.I.T., named Bernard Burke. Burke then heard about a recent seminar given by Jim Peebles, a young student at Princeton working in cosmology. Peebles had been studying a proposed radio antenna that his advisor Robert Dicke was building, which would search for the background radiation left over from the birth of the universe. Burke put the

The horn antenna used by Penzias and Wilson to first observe the 3 degree background radiation. Located in Crawford Hill, New Jersey, the antenna was designed for use with the Echo and Telstar communication satellite experiments.

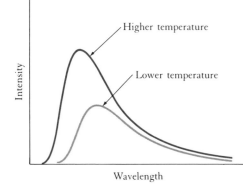

A *black body* is defined as one that absorbs all radiation that falls on it. It is not only a good absorber but a good emitter of light. The spectrum of emitted radiation depends on its temperature and has the characteristic shape shown in the graph above. A higher-temperature black body has a higher intensity at all wavelengths, and its peak emission is at a higher frequency (shorter wavelength) than that of a lower-temperature black body.

two groups together, and it was realized that Penzias and Wilson had indeed discovered the radiation that the Dicke group intended to find. A short time later, Dicke's group completed their antenna and verified that indeed the background radiation of the universe had been discovered.

According to the big-bang theory, the primordial high-density-high temperature plasma should have filled the universe with a sea of radiation corresponding to its temperature. The high temperatures would make radiation of extremely short wavelengths. As the universe expanded and cooled, the wavelength of the ambient radiation would grow longer until, at present, the wavelength, according to a calculation first worked out by Alpher and Hermann (following Gamow's ideas), should average a few millimeters in length. This radiation is characteristic of a black body "glowing" at a temperature of 3 degrees Kelvin, that is, 3 degrees above the absolute zero of temperature. It had been predicted as early as 1946 that a crucial test of the big-bang hypothesis was that the entire universe would be filled with such radiation.

This discovery illustrates several things. Scientists often make discoveries serendipitously, but whether the discovery is made or missed is crucially dependent on care, alertness, and imagination. In this case, the

The Arecibo radio telescope in Puerto Rico is a nonsteerable antenna incorporating an entire mountain. It observes objects that pass overhead as the Earth rotates.

earlier astronomers who used Bell Laboratories' antenna merely reset their zero and ignored the noise, thus missing the discovery and the Nobel Prize, which was awarded to Penzias and Wilson. Similarly, it should be noted that the Dicke group could have collaborated with the Bell Laboratories' scientists and used their existing antenna, but they wanted to try to do it all themselves, thereby also missing the Nobel Prize. There was also the irony of the theoretical work of the famed Soviet scientist Yakov Zel'dovich (coinventor of the Soviet hydrogen bomb with André Sakharov). Zel'dovich was aware of the predictions that if the universe had started out extremely hot and dense, there would now be remnant radiation of a few degrees. Zel'dovich was also aware of the Bell Laboratories' antenna and of the earlier publications coming from it, which did not report any background noise. But because Zel'dovich did not realize that American scientists might reset their zero, he concluded that the early universe must not have been hot and dense. It was only when the Penzias and Wilson discovery came out that it was clear what had happened.

Actually, there was still a problem: the spectrum of the background radiation was not known precisely. In 1973, balloon experiments by Berkeley physicist Paul Richards and his colleagues examined the details

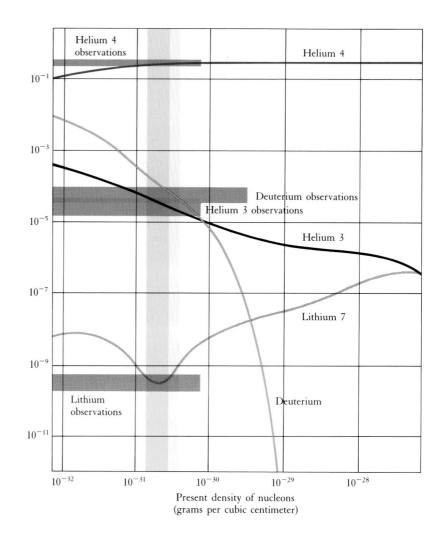

Predicted abundances of helium 4, helium 3, deuterium, and lithium 7 in the big-bang model of the universe (curves) closely agree with the observed abundances (shaded horizontal bands). The predicted abundances change as a function of the density of nucleons (protons and neutrons) at the time of the big bang; the shaded vertical band indicates the range in which all the observed abundances are consistent with the predictions. The fact that the predictions fit elements from lithium to helium in abundances ranging over more than a factor of a billion is one of the strongest arguments supporting the big-bang theory.

of the microwave part of the background radiation, and proved that the radiation really had the pure thermal spectrum of a black body as expected from the big bang.

In addition, measurements of the relative abundance of elements in external galaxies, done by astronomers from all over the world, confirmed the big-bang nucleosynthesis calculations that helium 4 represents about 25% of the mass of all the elements making up the universe. Furthermore, other light elements that were made in the big bang, for example, deuterium and helium 3 (at about 1 part in 100,000) and lithium 7 (at about 1 part in 10 billion) were found to have abundances exactly as predicted by big-bang nucleosynthesis, thus establishing that big-bang nucleosynthesis worked amazingly well. (We will return to this point later.) By the mid-1970s, it had been firmly established that we do live in a hot big-bang cosmology.

Subsequent work by balloon experiments and high-flying aircraft looked at the details of the background radiation. These measurements verify that the radiation is really of uniform intensity and isotropy, and there is no preferred direction. The measurements were accurate to a few parts per thousand.

INFRARED, ULTRAVIOLET, AND GAMMA RAYS

The other electromagnetic wavelength regions are not as easily accessible with ground instruments as are radio and optical waves. The Earth's atmosphere, transparent to visible light and radio waves, acts as a shield against the intense infrared and ultraviolet light, as well as against X rays and gamma rays. If it were not for this shielding, the incidence of skin cancer, not to mention genetic defects, would be considerably higher. Thus, to observe light from space in these other wavelength regions requires getting above the Earth's atmosphere. Astronomers began developing tools for doing just that during the 1960s. In particular, balloons were flown almost to the top of the atmosphere with sophisticated detectors attached, and rockets were shot up on short suborbital flights, carrying detection equipment above the atmosphere for a few brief moments. Even high-flying airplanes were used to get above the bulk of the atmosphere. Each of these early probes was tantalizing, revealing that there were indeed signals coming to us in other wavelength bands from objects in space.

For example, a clever rocket launch was able to show that X rays were coming to us from the Crab Nebula, a star that blew up in A.D. 1054. This was done by shooting the rocket up just as the Moon was passing in front of the Crab, so that the X rays were blocked by the Moon. However, it was not until satellites and space probes could stay up for long periods that astronomy in the nonoptical regions was able to come into its own. The Uhuru satellite, which flew in 1970 and provided a complete survey of the X-ray sky, found that many objects, particularly rather exotic ones such as neutron stars, supernova remnants, and black holes, emitted X-ray light.

The violently active galaxy known as Centaurus A photographed in different wavelengths. *Top*: An optical photograph from observations taken with the U.S. Cerro Tololo 4-meter telescope in Chile. *Middle*: A radio image from observations using the Very Large Array in New Mexico. *Bottom*: An X-ray image from observations made by NASA's Einstein satellite.

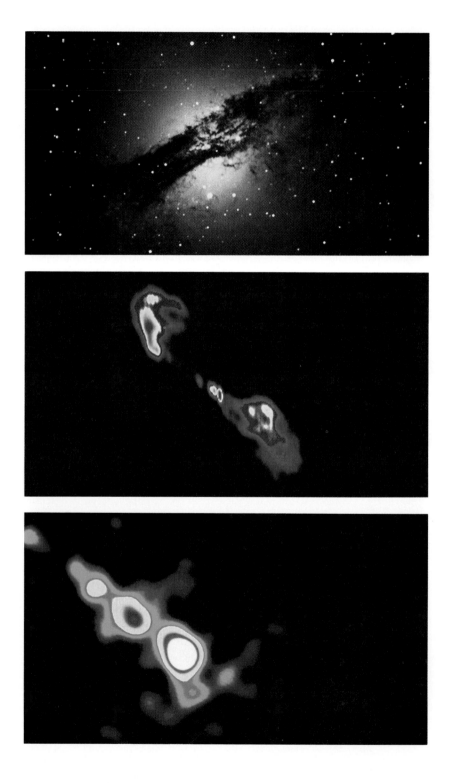

Similar stories of discovery can be made in the ultraviolet and infrared regions. There have also been some attempts at establishing gamma-ray observatories. All have found that there are indeed astronomical sources for each of these electromagnetic wave bands. Astronomy is not merely done with ground-based telescopes, but also, increasingly, with space observatories. We mentioned earlier the hope that an optical telescope will be put into orbit around the Earth, thus relieving the problem of looking through the Earth's atmosphere, even in the optical light. This instrument, the Hubble Space Telescope, will manage to avoid the distortions introduced by the atmosphere, which have limited the quality of Earth-based observations.

NEUTRINOS AND THE SUPERNOVA

In addition to telescopes looking at all the electromagnetic radiation bands, a new kind of telescope, and thus a new kind of astronomy, has recently been established. This is one that directly ties itself to elementary particle physics and illustrates another example of the inner-space/outer-space connection. As we saw in Chapter 3, neutrinos were postulated from the properties of beta decay, and were eventually found at reactors and subsequently at accelerators using very large detectors. When large detectors are placed deep underground to shield them from the cosmic rays, the only things that can get down to them are neutrinos and, occasionally, extremely high energy muons. All other particles are shielded from the detector by the layers of earth, often as much as a mile thick. These underground detectors were built in the early 1980s to look for the possible decay of protons, and they could tell the difference between the decay of a proton in the detector and the burst of energy characteristic of the collision of an incident neutrino with a nucleus of detector material. Some steady, low-rate background "noise," due to natural radioactivity and to cosmic-ray-produced neutrinos and muons, was to be expected, but any sort of burst observed in the detector would imply some interesting astronomical neutrino activity.

Just such a burst of background neutrinos was detected on February 23, 1987, as we saw in the story at the beginning of Chapter 1. These neutrinos preceded the flare of light from a supernova that was just reaching Earth after its explosion in the Large Magellanic Cloud 170,000 years ago. Thus, the Kamioka and Lake Erie detectors established neutrino astronomy, and were able to observe a supernova radiation of neutrinos in amounts that were completely consistent with what was expected.

We have seen that astronomers can study outer space not only with optical light but with radio, microwave, infrared, ultraviolet, X rays, and gamma rays, and now by means of neutrinos. Each of these different ways of observing space has yielded new and interesting information, and has caused a revolution in our way of thinking about objects in space.

THE DEUTERIUM STORY: OPEN VERSUS CLOSED UNIVERSE

In addition to the abundance of helium 4 mentioned earlier as confirming the big-bang theory, other light elements also produced in the big bang have given us quantitative information about specific properties of our universe. Here we will show how the quantity of deuterium in the present universe gives us an important clue about its future evolution. Deuterium is a form of the element hydrogen, which has a single proton as its nucleus. The nucleus of the deuterium atom, however, also contains a neutron, giving it an atomic weight of 2 instead of hydrogen's 1. Deuterium was made only in trace amounts in the big bang, and it turns out that that trace is a very steep function of the density of the newly born universe. At higher densities, all the deuterium is destroyed and converted to helium 4; at low densities, lesser amounts of deuterium are destroyed by nuclear collisions, leaving some deuterium to remain to the present day. Deuterium has a very fragile nucleus; it doesn't take much jostling to separate the neutron from the proton. In the early 1970s it was shown that any stellar environment would end up destroying deuterium rather than producing it. Thus, any deuterium seen today is directly traceable to pristine material from the big bang.

Significant deuterium abundance determinations were first made during the Apollo-11 Moon mission. On this mission, displayed to millions of TV viewers, astronauts pulled out an aluminum foil window shade which they exposed to the solar wind. The solar wind is a flux of particles, mostly protons, which escape from the surface of the Sun and spread throughout the solar system. This spray of particles embedded itself in the aluminum foil, which was rolled up again, and taken later to a terrestrial laboratory in Bern, Switzerland. There, it was analyzed by Johanus Geiss, who inferred that the material from which the sun was formed had deuterium to the order of 1 part in 100,000. This discovery was subsequently verified in 1973 when the Copernicus satellite measured the ultraviolet lines of deuterium, and compared them with the

Astronauts Neil Armstrong and Edwin Aldrin deploying the aluminum foil "window shade" during the Apollo 11 lunar landing in July of 1969. The solar wind particles collected by the window shade yielded evidence about the abundance of light elements in the sun.

ultraviolet lines of hydrogen to establish a deuterium-to-hydrogen ratio for gas in the interstellar medium. It was found that the deuterium–hydrogen ratio was in almost exact agreement with the numbers implied by Geiss' solar-wind experiment, and that the deuterium really existed at a level of about 1 part in 100,000.

Back in the 1960s, when the first detailed calculations of the yields implied by big-bang nucleosynthesis were being made (by William Fowler at Cal. Tech., in collaboration with Fred Hoyle and Robert Wagoner, and by Jim Peebles at Princeton), it was shown that helium 4 always accounted for about 25% of the mass, and that traces of the other light elements were made. At that time, it was generally thought that the traces of other light elements were irrelevant, because they could also be made during the process of star formation. However, in 1969, Hubert Reeves, a French-Canadian living in Paris (who is now a radio and television personality in France, somewhat analogous to Carl Sagan in the United States), showed that the Fowler–Hoyle model for making light elements during star formation failed, because it required more energy than was available. Subsequent work by David Schramm and his collaborators showed that, as mentioned before, the nuclear physics of

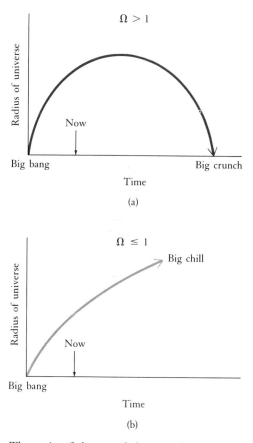

The ratio of the actual density of the universe to the critical density is called Ω. (a) If Ω is greater than 1 (density greater than the critical value), the universe will eventually contract into the big crunch. (b) If Ω is less than or equal to 1, the universe will continue to expand into the big chill.

deuterium prevent it from being formed in any stellar situation, and that the big bang was the only thermodynamic environment in which deuterium could be produced and survive. Thus, if we can measure the abundance of deuterium, we have a direct probe into the big bang. This enabled the observations of Geiss and the Copernicus-satellite group to be used to pinpoint the conditions in the big bang that were necessary to make that amount of deuterium. The conclusion of Reeves and Schramm and their collaborators was that the density of the material entering the nuclear reactions that made deuterium in the early universe was only about 10% of the so-called critical density of the universe.

The *critical density* is the boundary that separates the possible future histories of the universe (see figures on this page). If the density is above its critical value, the gravitational pull of the universe upon itself will eventually stop the universe's current outward expansion and cause it to contract. This is known as the *closed universe* prediction. But if the density of the universe is below the critical point, the universe will continue to expand forever; this is known as an *open universe.*

For a comparison, imagine shooting a rocket off from Earth (see figure on the facing page). If the rocket is shot with enough velocity, it will continue to move away from the Earth as an interplanetary probe. However, if it does not have enough initial velocity, even though it goes up at first, it eventually falls back down in a ballistic trajectory. The boundary between the two—the critical velocity—is where the rocket goes up and has just enough velocity to go into orbit as a satellite. For the universe, we have the expansion rate as determined by the Hubble constant. The question is what gravitational pull the universe has upon itself: Can it escape its own gravity?

The deuterium argument told us that all the normal matter that enters into nuclear reactions could provide only 10% of the gravitational pull required to stop the expansion. Subsequent work of a similar nature by theoretical astrophysicists at the University of Chicago and their collaborators showed that the same kind of arguments could be developed for the other light elements produced in the big bang: lithium 7 and helium 3. These elements also pointed to a low density of normal matter. Hence, there is not enough normal matter to close the universe. If the universe is made out of only normal matter, it will continue to expand forever. Furthermore, the amount of matter implied by these trace-element studies, though not enough to close the universe, was more than the amount of matter that could be estimated by counting stars. This indicated that some of the matter in the universe is not shining. Observations of material in orbit around galaxies also show that strong gravitational forces are being exerted by nonshining matter. This mysterious stuff is called *dark matter,* and there is at least ten times as much of it as there is of shining matter. If there is enough nonshining matter to make

We can predict whether a rocket will fall back to Earth or continue into space because we know both the velocity of the rocket and the gravitational field of the Earth. We can use the Hubble expansion to estimate the velocity at which the universe is expanding, but the fate of the universe remains unknown because we do not know its density and hence its gravitational field.

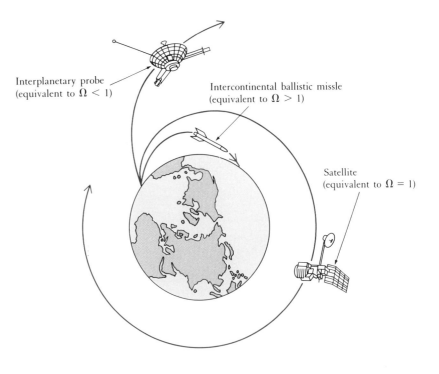

Interplanetary probe
(equivalent to $\Omega < 1$)

Intercontinental ballistic missle
(equivalent to $\Omega > 1$)

Satellite
(equivalent to $\Omega = 1$)

the density of the universe equal to the critical value, then, from what we know about how matter was synthesized during the big bang, we know that some of that nonshining matter would have to be something other than the normal kind of matter that enters into nuclear reactions and that is made up of neutrons and protons. But nucleosynthesis also tells us that some of the nonshining matter is just ordinary neutrons and protons that aren't in stars.

THE BIRTH OF THE INNER-SPACE/ OUTER-SPACE CONNECTION

With the establishment of the big bang, it became clear that the early universe was in some sense an elementary particle-physics laboratory, since the energies and densities in the very early universe were enormous, far greater than even those achieved in particle accelerators. In fact, the physics governing what was going on in the early universe is the physics of elementary particles.

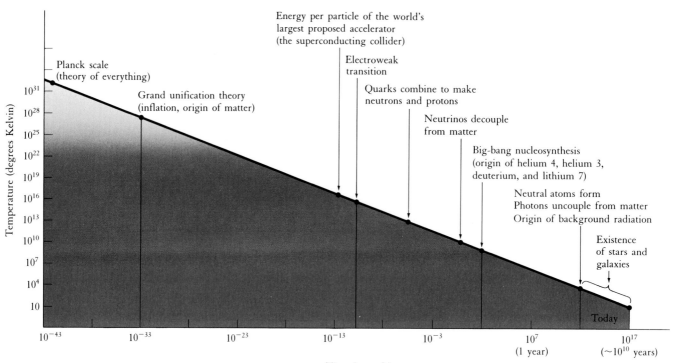

Energy per particle of the world's
largest proposed accelerator
(the superconducting collider)

Electroweak
transition

Quarks combine to make
neutrons and protons

Neutrinos decouple
from matter

Big-bang nucleosynthesis
(origin of helium 4, helium 3,
deuterium, and lithium 7)

Planck scale
(theory of everything)

Grand unification theory
(inflation, origin of matter)

Neutral atoms form
Photons uncouple from matter
Origin of background radiation

Existence
of stars and
galaxies

Today

Time (seconds)

The thermal history of the universe, starting 10^{-43} seconds after the big bang and continuing to the present. Note that all the predicted interactions of elementary particle theory such as grand unified theories and the theory of everything had an epoch when they dominated the physics of the early universe.

From the time of the first instants of the big bang, the universe has passed through an enormous range of temperatures, from something exceeding 10^{32} degrees Kelvin to our present average temperature of 3 degrees Kelvin.

Temperature is a measure of the average kinetic energy of the particles. Thus, we can relate these temperatures to particle energies. In our favorite units (electron volts), 1 eV is equal to 10^4 degrees Kelvin, or 10^4 K, (units of thermodynamic temperature). This means that in a gas heated to 10^4 degrees Kelvin, the molecules whiz around with an average energy of 1 eV. At this temperature (or energy) the molecular collisions are violent enough to disrupt such molecules as H_2O or N_2, breaking them up into their constituent atoms. What does the universe look like at, say, 10^6 degrees Kelvin? This is equivalent to 100 eV, and here the collisions between atoms are strong enough to strip away the outer electrons, since their binding energies are less than 100 eV. The state of the early universe at higher temperatures is shown in the figure on this page and in the table on the facing page.

Our earthly accelerators, by dint of tremendous ingenuity and expense, can organize some 10^{12} or 10^{13} (less than one billionth of a gram) particles per second to make the kinds of collisions that all the particles

··

EVENTS DURING THE EARLY BIG-BANG UNIVERSE

Time (t)	Temperature* and energy	Events as the universe expands and cools
$t \sim 10^{-43}$ sec	$\sim 10^{32}$ K (10^{19} GeV)	Earliest time that physics as we know it can exist. Time and space no longer have their normal meaning at this epoch. We cannot extrapolate to $t = 0$; time has no meaning in units of less than 10^{-43} sec. That is, 10^{-43} sec is the smallest unit into which we can subdivide our concept of time.
$t \sim 10^{-6}$ sec	$\sim 10^{13}$ K (1 GeV)	The universe cooled down to temperatures comparable to those inside an atomic nucleus. The quark soup of the early universe condensed into baryons and mesons; the baryons being clusters of three quarks and the mesons being quark–antiquark pairs.
$t \sim 1$ sec	$\sim 10^{10}$ K (1 MeV)	Neutrinos cease interacting in any significant way with the rest of the universe.
$t \sim 3$ min	$\sim 10^{9}$ K (100 keV)	Nuclear reactions take place: deuterium, helium 3, helium 4, and lithium are synthesized.
$t \sim 10^{5}$ yrs	$\sim 10^{4}$ K (1 eV)	Plasma of nuclei and electrons begins to condense into atoms. Photons can no longer interact with matter; they decouple and begin propagating freely (as they have ever since; they are now the 3-degree background radiation). Universe is now dominated by matter, not radiation: matter begins condensing under its own gravity into galaxies, stars, etc.

*The relationship is: At 300 degrees Kelvin, or 300 K, (room temperature) the average energy of molecules is one thirtieth of an eV.

in the universe participated in during the earliest instants after the big bang. This is a fundamental connection between inner and outer space.

What actually established the robustness of the connection of these two fields was the realization that the flow of information went both ways. Astronomy changed from being a parasite on physics, using the physics measured in laboratories and applying it to stellar situations, to being able to make predictions about fundamental physics, telling physicists things as yet unmeasured in the laboratory.

A significant example of this feedback came with regard to the number of elementary particles. The big-bang theory implied the mass fraction of helium in the universe. That mass fraction, 25%, has a very slight dependence on the number of types of neutrinos. As we have seen, each neutrino is associated with a family of elementary particles.

Thus, more neutrinos, if this connection persists, would imply more families of elementary particles. In the mid-1970s, it appeared that whenever physicists explored higher energies, new kinds of particles were discovered, and for all we knew, families of elementary particles could proliferate indefinitely.

However, theoretical astrophysicists showed that the amount of helium observed was consistent with the existence of only a few types of neutrinos and, therefore, a few families of elementary particles. Subsequent work has refined this prediction, stating that the total number of neutrinos (and thus the total number of elementary particle families) cannot exceed four, and probably fits best with only three families (see the figure on this page). This was a radical statement, because it implied that astronomical observations, coupled with cosmological arguments, could predict something as fundamental to the nature of matter as the number of elementary particles.

The original prediction about the number of particle families was made in the mid-1970s. By the end of the 1980s, we have begun to check these conclusions in the laboratory. For the first time we are in a position

The abundance of helium 4 suggests there are at most four families of elementary particles. The three curves represent an enlargement of the part of the helium-4 curve lying within the shaded vertical band in the illustration on page 144; the narrow curve in that illustration resolves into the three broad curves. The bottom curve shows the helium-4 abundance predicted if there were two families of particles. The middle curve shows the abundance predicted for three families, and the top curve shows the abundance predicted if there are four families. The predicted abundances of helium 4 for two and three families of particles are well within the region defined by helium-4 observations and estimates of nucleon density (green region). A fourth family would produce an abundance that would be very close to the allowed extremes.

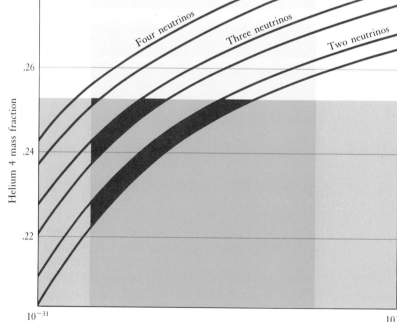

where laboratory ultrahigh energy accelerator experiments are being used to check cosmological predictions, thus changing cosmology into a laboratory science.

The early tests from accelerators such as those at CERN and Fermilab indicate that the number of neutrino species is small. These tests are done by producing the intermediate-vector boson Z^0, which can decay into any neutrino species.

For example, one decay mode that must exist is:

$$Z^0 \rightarrow \nu_e + \overline{\nu}_e$$

The Z^0 decays into an electron neutrino and its antiparticle. For each neutrino species, there is a similar pair of particles into which the Z^0 can decay. Each such pair increases the probability that the Z^0 will decay; so the more neutrino species there are, the faster the Z^0 decays. By measuring the lifetime of the Z^0, we can calculate a limit on the number of neutrino types. That limit is already quite small. As new accelerators are built, such as the Stanford Linear Collider, and as new experiments are conducted at the Tevatron at Fermilab and at CERN, this laboratory limit on the number of neutrinos should be brought down to what has been predicted by cosmologists.

In fact, a completely independent verification of this number came from the supernova that appeared on February 23, 1987. The neutrino detection from the supernova was most sensitive to one specific kind of neutrino, the antielectron neutrino. Because the energy of the supernova was radiated in all kinds of neutrinos, there could not have been very many kinds of them, or else the percentage of them radiated in the detectable species would have been reduced below the limits of what the equipment could detect. The limit calculated from the supernova data tells us that there are no more than seven species of neutrinos, a number that is at least comparable to the limit of five or six species that has been obtained from laboratory experiments up to 1989. Thus, it appears that particle physics and cosmology are indeed intimately related, and that the predictions of one area can be verified by experiments in the other.

THE STANDARD MODEL OF BIG-BANG COSMOLOGY: PROBLEMS AND ISSUES

We have seen that the standard big-bang model, with the early universe being hot and dense, is now reasonably well established. It has made detailed predictions about abundances of elements, ranging over nine orders of magnitude, that agree with observations. It has even made predictions about such fundamental things as the number of elementary particles—predictions which today at least appear consistent with observations. However, there remain certain questions.

One question is whether the universe is going to expand forever or collapse back on itself. Big-bang nucleosynthesis tells us that the quantity of normal matter is not enough to create sufficient gravitational pull to stop the universe's expansion. But could there be some other form of matter, some exotic particle, for example, which adds to the normal matter and brings the universe up to its critical density? The observations of the dynamics of the universe are not yet sensitive enough to resolve directly the future of the universe, nor have the mystery particles been found. We will discuss the experiments and instruments planned to investigate this problem in the next two chapters.

An area of great concern, once it was established that we live in a big-bang universe, is the initial conditions for the big bang. For example, where did our space of three spatial dimensions plus one time dimension come from? What caused the expansion of that space? Why is the universe so smooth on some scales, yet so lumpy on others? How did the universe have matter in it in the first place? In the laboratory, whenever we make new matter, we always have to make an equal amount of antimatter. Yet the universe does not appear to have any large concentrations of antimatter or any boundary between matter and antimatter where annihilation is occurring. It simply seems to be made out of matter. How did that matter get formed? In the next two chapters we will also be discussing the instruments and experiments being designed to probe these initial-condition problems.

One of the major mysteries of modern cosmology is how the smooth early universe was able to fragment and break up into the system that we see today with clumps of galaxies, stars, and people. Clearly, some seeds for the formations of these clumps had to have existed in the early universe. The origin of those seeds is thought to be related to particle phenomena in the early universe. This is again one of the potential connections between cosmology and particle physics.

SOME SUCCESSES OF THE BIG-BANG THEORY

As we have seen, the big bang has enabled us to explain the background radiation and the light-element abundances. The idea of an evolving universe helped us explain why quasars are primarily seen at large distances, since they are assumed to be associated with galaxy formation. Galaxy formation occurred relatively early in the history of the universe. Because light takes a finite amount of time to travel long distances, by looking at objects from great distances, we see not how the objects are now but how they were in early times. There is also the interesting fact that all the different arguments about the universe's age give approximately the same result. The dynamics of the universe using the Hubble age gives an estimate of about 10 to 25 billion years. The ages of the oldest stars range from 12 to 18 billion years, as does the age implied from radioactive dating of the heavy elements. All this evidence is completely consistent at around 15 billion years, given the accuracies of the different techniques. These totally independent methods all give the same age. Therefore we have another independent verification that something interesting and exciting did happen about 15 billion years ago. All of these confirmations of the big bang tell us that the early universe was, indeed, hot and dense. By extrapolation, it is clear that matter was once so hot that it was decomposed into its most primordial constituents, quarks and leptons. Thus, the physics of high energy dominated the very early universe, and perhaps the only way to explore the fundamental physics of this early universe in a controlled way is in particle accelerators.

The observations done with optical telescopes to establish the universe's expansion, those with radio and microwave telescopes to find the background radiation, those with satellite-carried telescopes, even those with the Apollo Moon mission to establish the abundances of the elements, as well as the various age arguments, all confirm that the big-bang idea seems to be right. Accelerators are now checking these ideas by verifying the number of elementary particles to see whether the big-bang predictions were correct. The two fields have become quite symbiotic. Predictions from one field are checked by measurements in the other. These predictions help explain mysteries and paradoxes in the interrelated fields.

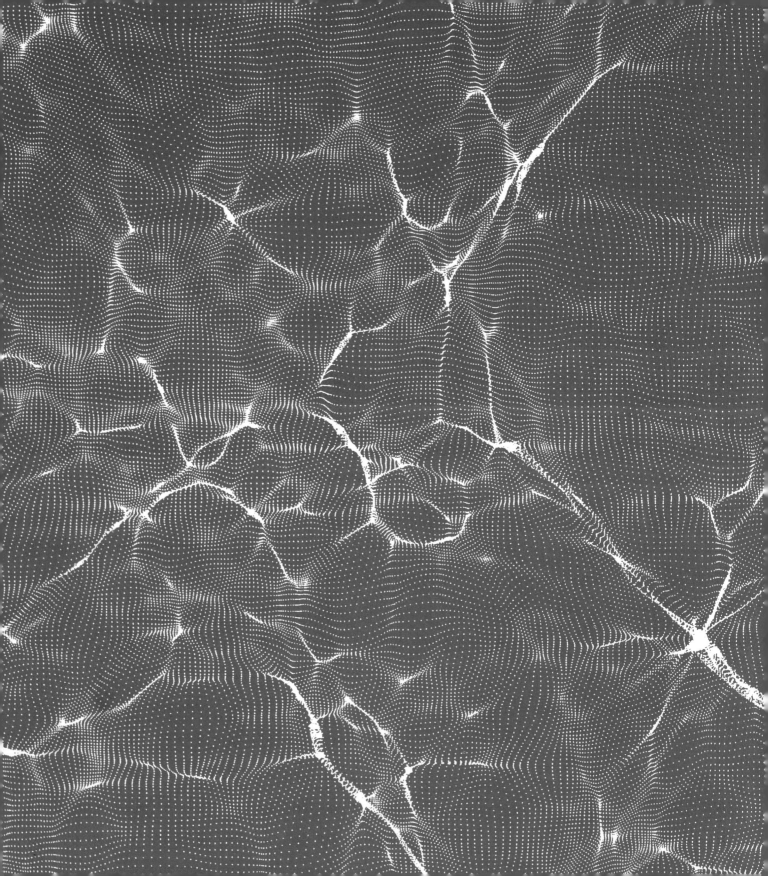

6 THE CONVERGENCE OF INNER SPACE AND OUTER SPACE

In Chapter 5 we saw that the convergence between inner and outer space began to accelerate when big-bang nucleosynthesis made predictions about the number of fundamental particle types. Convergence of the disciplines was also served when particle physics clarified some long-standing problems in cosmology, so that information was flowing in both directions.

In this chapter we will see how the new developments in particle physics, particularly the attempts to create a grand unification theory, have solved some cosmology problems, and how these particle-physics theories may be tested by means of telescopes as well as accelerators. Let us look at these theories, and explore some of the language that physicists use in groping for a more complete understanding of the nature of matter, space, and time.

GRAND UNIFIED THEORIES

We begin by surveying the ways in which the theorists are attempting to extend the standard model. An underlying mathematical property characterizing each of the fundamental forces is a special symmetry mentioned earlier—the gauge symmetry. In our discussion of symmetry in Chapter 4, we noted that symmetry is connected to conservation laws; in the case of electromagnetism the gauge symmetry is connected to the well-established law of conservation of electric charge. The word *gauge* is used because the symmetry is that of scale or size, with scale being like

A computer simulation of the growth of structure in the universe.

the "gauge" used to describe scale in model railroads. If a force is carried by a particle that has zero mass, then, in principle, the force can act over any distance, and so has no length scale attached to it; hence a massless particle, such as a photon, can travel forever. Thus, the particles that carry forces subject to gauge symmetry must have zero mass. They are called gauge particles or gauge bosons.

However, as we have seen, the W and Z particles that carry the weak force are not massless at the low energies at which we live, but have masses of about 100 GeV. For this reason we say that their gauge symmetry is "broken" at low energies and restored at high energies, where the Z^0 becomes like an ordinary photon. Explaining the gauge transformations allowed Weinberg and Salam to unite the nuclear weak interaction and electromagnetism. Weinberg and Salam could imagine a situation in space-time where the temperature, or average energy, of particles in some region in space, such as the collision region of an accelerator, was so extreme that the two interactions were of equal intensity.

This same principle is applied in attempts to bring the nuclear strong interaction into a synthesis with the electroweak force. These unification theories are known as "GUTs," for grand unification theories. Unifying the electroweak and the strong force would require energy conditions about 10^{15} times more intense than those necessary to generate protons. To unify these with the final force, gravity, into a theory of everything (TOE) would require even higher energies, estimated to be about 10^{19} GeV. To build an accelerator that achieves 10^{15} GeV, with the technology of the Tevatron, would require a machine so large as to stretch to the nearby stars. To achieve 10^{19} GeV would require galactic scales. Obviously we will not be able to build such an accelerator in the near future.

Enormous energies are required to unify the four forces of nature. It is thought the four forces were once unified at the higher energies characteristic of the universe soon after the big bang, and indeed a theory that unifies the weak force and the electromagnetic force has already been verified for energies of a few hundred GeV.

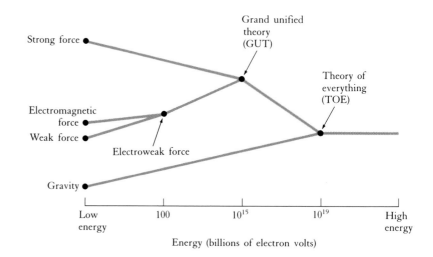

Energy (billions of electron volts)

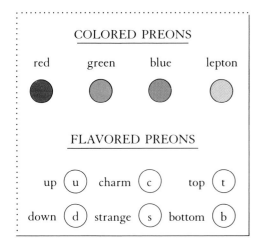

COLORED PREONS

red   green   blue   lepton

FLAVORED PREONS

up (u)   charm (c)   top (t)

down (d)   strange (s)   bottom (b)

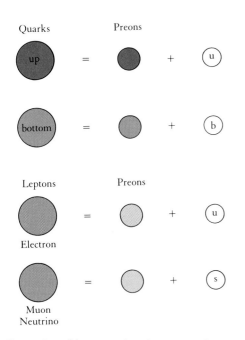

Quarks   Preons

up = + (u)

bottom = + (b)

Leptons   Preons

= + (u)

Electron

= + (s)

Muon
Neutrino

Examples of how a colored preon and a flavored preon could be combined to form quarks and leptons.

However, we know that, because of the big bang, the universe itself went through a period in which it had such energy densities, and in which these forces were theoretically unified. One way to test these unification theories is to see if they have any implications for cosmology. It is this interplay that has brought cosmology and particle physics together in studying the very early universe. And so, back to particles.

All our experiments indicate that quarks and gauge bosons interact as points with no spatial dimensions, and so are fundamental, like the leptons. If the fundamental particles really are dimensionless points with mass, flavor, color, charge, and other quantum properties, occupying no volume, then the nature of matter appears quite bizarre. *The four interactions give matter shape; matter itself is empty.* If all four interactions really are unified, and the particles themselves are just points of interaction for the force, then the phrase from George Lucas's *Star Wars* movies, "May the force be with you," might really be appropriate.

However, as we reduce the number of forces by unification, we keep increasing the number of particles. This situation has prompted some theorists, such as Abdus Salam, to suggest that there is another level of particle reality below quarks. Salam calls them *preons* and argues that each quark consists of two preons, one defining flavor, the other defining color. Salam brings leptons into the picture as well by assigning them a fourth color. With this scheme, all the leptons and quarks can be constructed from just ten preons (see the margin of this page).

Salam's speculative idea points out that the old problem of what is really fundamental may not yet be fully resolved. The line of reasoning guiding modern theoretical physics can be expressed as: "Underneath it all, isn't there just *one* force that produces all the particles and their interactions?"

When we try to extend GUTs into a TOE, to bring gravity into the picture, we have a new problem. Although there is a nice match-up between quarks and leptons, there being six quark flavors and six leptons, there is no such connection between matter-making fermions and the force-carrying bosons, and yet gravity interacts equally with both fermions and bosons. Thus, if we want a true theory of everything, we must have some way of changing fermions to bosons and vice versa. The theoretical solution is called supersymmetry (or SUSY for short). SUSY postulates that there is an entirely new kind of matter that creates a symmetry between the fermions and the bosons—that each of the presently known fermions has a boson partner, and each of the known bosons has a fermion partner. Thus, the number of known particles would be doubled. If this theory is true, this new kind of matter should eventually be seen in accelerators. The fact that none of these particles have yet been discovered (as of 1989) indicates that the lightest of them is heavy, and therefore difficult to make in current accelerators. However, the best predictions are that SUSY particles should appear just above the masses of the heaviest standard-matter particles, the W's and Z. Therefore, the next generation of experiments might discover a whole new realm of matter. We may not need to build accelerators that reach 10^{19} GeV to see indications of a TOE.

Let us now see how GUTs and/or SUSY might find applications in our understanding of the birth of the universe.

THE ORIGIN OF MATTER

Our primary source of information on the structure of the universe comes to us in the form of radiant energy, with which the universe is awash. In the previous chapter, it was shown that the big bang controlled the production of chemical elements, congealed out of the quarks as the primal fires cooled. This theory of big-bang nucleosynthesis insists that the observed abundances of the isotopes of hydrogen, helium, and lithium establish limits on what the density of normal matter in the universe may be. Another method for calculating the amount of matter in the universe relies on measuring the velocity of light-emitting things—atoms and stars, whose orbits completely surround galaxies. This rotational velocity, according to Kepler's third law as derived by Newton, depends on the total mass contained inside the orbit; so the mass of the galaxy can be calculated. Much to the surprise of observers, the amount of mass continued to increase even for stars and atoms orbiting far outside the luminous region of the galaxy. Some nonluminous, or dark, matter was pulling on these atoms. (We will return to the dark-matter problem later.)

A useful quantity in our discussion of the origin of matter is the ratio of matter to light. There are about 10 billion photons for every proton or neutron in the universe. How did the universe come to have this ratio of matter to radiation? Until the late 1970s, the only possible answer to this question was just, "In the beginning there was just one proton produced to every 10 billion photons." Now with the guidance of grand unified theories, we have some good ideas about how this ratio might have been generated. In essence, although matter and energy are completely interchangeable at a high enough temperature, as the universe cooled, the now lower energy photons could no longer convert directly into baryons.

THE SEARCH FOR ANTIMATTER

In the laboratory, whenever we produce a particle out of pure energy, we also produce its antiparticle. As a result, some physicists supposed that the universe must originally have been equally divided be-

The final inflation stage of a NASA-sponsored heavy lift balloon designed to carry scientific instruments up to an altitude of 30 miles. The balloon is made of a thin, polyethylene material 0.8 millimeters thick. At full inflation, it will reach 30 million feet in volume.

tween matter and antimatter. Our local system is made of matter. The question that then tormented the astronomers was: Where is the antimatter?

However, in the late 1960s, Gary Steigman (now of Ohio State University) and Yakov Zel'dovich showed that in any universe that was symmetric in matter and antimatter, protons and antiprotons would continue to annihilate each other until there was only one proton or antiproton for every 10^{18} (10 billion billion) photons. Thus, the present ratio of one proton to 10 billion photons argues against such a symmetric model.

Furthermore, we can search directly for antimatter by looking for it in cosmic rays, which come from deep space. In fact, the most energetic particles travel so rapidly that they are not even confined to galaxies. Observations near the top of the atmosphere in balloons and rockets have found that the ratio of antiprotons to protons in the cosmic rays is about 1 to 10,000, which is the amount of antimatter we expect to be generated by proton collisions with the material of the galaxy, judging from collisions at accelerators. It therefore does not appear that there are any extra antiprotons anywhere in the vicinity of our galaxy.

Another search technique makes use of the fact that matter–antimatter annihilation will produce gamma rays of a certain characteristic energy. Since the boundary between regions of matter and antimatter would generate huge amounts of annihilation, we should be able to find such a boundary, if it exists, by searching for these characteristic gamma rays. Searches carried out by gamma-ray detectors in satellites have failed to observe any significant sources; so, as far as we can tell, our region of space is all made of matter. Thus, we are presented with a conundrum: If matter–antimatter symmetry holds, why does the universe seem so asymmetric?

NONCONSERVATION OF BARYONS

All the current GUTs predict that, at a high enough temperature, the number of quarks, and hence the number of neutrons and protons, is no longer kept constant; that is, in grand unification theories, baryons are not conserved. Allowing the total number of quarks to change means that protons can evaporate or decay away. The normal strong, weak, and electromagnetic forces do not change the total number of quarks, but when they are combined into a unified force, the theory insists that this is no longer true. The unified force can change quarks into leptons and vice versa. For quarks and leptons to be on equal footing, they must be able to easily change into one another. In the early, hot universe, where the symmetry was perfect, there must have existed exchange particles that changed quarks and leptons into one another. Enter the conjectured

X and Y bosons, each with a mass 10^{15} times that of a proton, or about 10^{15} GeV, generated by a Higgs field similar to the one that generated the W and Z masses but at a much higher energy.

At present accelerator energies (hundreds of GeV), quark nonconservation processes are almost negligible (*almost* but not totally, since GUTs do predict proton decay with lifetimes of about 10^{32} years). However, at particle energies approaching the mass of the X boson, these processes will be occurring as rapidly as all the other interactions. Energies of the order of 10^{15} GeV (which is called the grand unification energy) occurred in the very early universe, when the temperature was of the order of 10^{28} degrees Kelvin. In the standard big-bang model, this corresponds to a time about 10^{-35} seconds after the birth of the universe, when its density was enormous, and its observable size was minuscule. Particles and radiation were in violent interactions with one another.

Since the quark nonconservation interactions were so strong at that very early time, is it possible that a universe in which matter and antimatter were originally symmetrical (zero net quark number) might evolve into today's universe where there is an excess of quarks over antiquarks? For this to happen, two additional ingredients were necessary.

The first was particle–antiparticle asymmetry, or charge conjugation (C) and charge conjugation–parity (CP) violations. As we mentioned in Chapter 4, experiments with accelerators in the 1950s and 1960s showed that matter and antimatter are not completely symmetric. Tiny deviations from perfect symmetry showed up in the weak forces that govern the decay of pions, muons, and kaons. We call these breakdowns C and CP violation.

A CP violation would permit X bosons to decay into quarks or into antiquarks at slightly different rates. Hence this ingredient provides an indication of what would be preferred—matter or antimatter. Obviously, in our universe, matter has won.

Even with this, we still need another ingredient to create a net long-term excess of matter, because at high enough temperatures, equilibrium occurs, and reverse reactions will wipe out any excess. (Reverse reactions would be quarks combining to make X bosons.) Therefore, the second needed ingredient was a departure from thermal equilibrium at some point in the early universe. Such departures occurred naturally as the universe expanded and cooled.

Once the temperature dropped low enough, the quarks could not recombine anymore to make X's, but X's previously formed could still change into the lower-mass quarks. Therefore, the reactions could now only go one way, and are said to be out of equilibrium. The net result was that an initially symmetric universe can generate a slight excess of quarks over antiquarks, of about 1 in 10 billion. At that instant, 10^{-35} seconds after the universe's birth, the number of quarks was roughly equal to the number of photons. The tiny matter–antimatter asymmetry continued until quarks began coalescing into baryons, and antiquarks into antibaryons. These particles of matter and antimatter annihilated each other and generated photons, which, because of the cooling (then down to a "mere" 10^{12} degrees Kelvin), could not regenerate baryons. The small excess generated at 10^{28} degrees Kelvin left about 1 neutron or proton per 10 billion or so photons, as we see today. The tiny initial surplus of quarks over antiquarks thus accounts for all the quarks in the universe, since each antiquark "took out" a quark. Were it not for the infinitesimal initial asymmetry, the universe would consist entirely of radiation. Our local galactic cluster, our Milky Way Galaxy, our solar system, and Earth and the life that inhabits it would not exist!

To summarize, we can understand the origin of matter and the fact that we live in a universe made of matter, even though the laws of physics show strong matter–antimatter symmetry. We can even estimate correctly the ratio of matter to radiant energy. To do all this, we need the following:

1. A GUT that can change quarks to leptons and leptons to quarks, that is, baryon nonconservation.
2. A GUT that has some kind of asymmetry between particle and antiparticle reactions or decay rates. The Fitch–Cronin experiment verifies that such processes exist.
3. A universe that cools sufficiently that the reactions described in 1. and 2. drop out of equilibrium.

In order to have this grand unified theory work, the universe must have been hotter than about 10^{15} GeV (10^{28} degrees Kelvin), but the standard big-bang theory probably does not run into trouble until 10^{32} degrees Kelvin, for reasons that we shall see later. So it looks as if the GUT does work. (These basic arguments were made by André Sakharov in 1967 but no grand unified theories existed at that time. The arguments were rediscovered in the late 1970s after GUTs became common.)

THE INITIAL CONDITIONS OF THE UNIVERSE

We have just seen how the new grand unified theories of particle physics have been able to solve one of the longest-standing puzzles in cosmology, namely, the origin of matter. Let us now see how some of these new ideas can solve several other persistent problems, particularly those having to do with the initial conditions of the universe. How did the universe get to be so smooth, smooth on the largest scales, and yet be lumpy, that is, clustered into stars, galaxies, and globs of galaxies, on a smaller scale? How did the universe get to be so old? In addition, we might be able to solve a new problem that is introduced with the production of matter—the monopole problem.

The solution to all these problems may be one very special process that occurs because of general relativity on the one hand and grand unification of the forces of particle physics on the other. The process is called *inflation*. This is not the economic inflation that governments grapple with, but the cosmological inflation of the very early universe— an expansion much, much faster than the normal big-bang expansion that is going on today.

We will first describe the particular cosmological problems that inflation appears to solve. We will then show how inflation developed, what it is, and what is necessary to make an inflation model work. We will find out that the models that work seem to end up with our universe being a single bubble of space. There could be many, many other bubble universes existing out there, all out of communication with our universe. Thus, not only is the universe a very big place, but our universe might not be the only one.

Let us now begin with a survey of the problems that inflation addresses.

THE SMOOTHNESS PROBLEM

In Chapter 5, we described the relic 3-degree background radiation. We saw that this radiation seems to be uniform, coming with equal intensity from all directions. (Some of the most precise experiments using radio telescopes actually compare the strengths of signals at a given frequency from different parts of the sky. Comparisons are always more precise than absolute measurements.)

What we find is that the 3-degree radiation is remarkably smooth: Once the Earth's motion through the radiation is accounted for the radiation from all directions has the same temperature to an accuracy of better than 1 part in 10,000. Furthermore, it is smooth across distances so large that light could not have traveled between those points within the age of the universe. For example, if we look in one direction out to the edge of observability and then look in the opposite direction, we see the background radiation at the same temperature of 3 degrees. In our experience, things come to be the same temperature by being in contact, that is, having collisions; yet the radiation coming from one direction has

The cones represent the area reached by radiation in a given length of time. An observer today receives microwave signals from sources that were separated by 10 billion light-years at the time of emission. At that time, the universe was only 10,000 years old, one-hundredth of the travel time between the sources; thus, the two sources could not have come into contact.

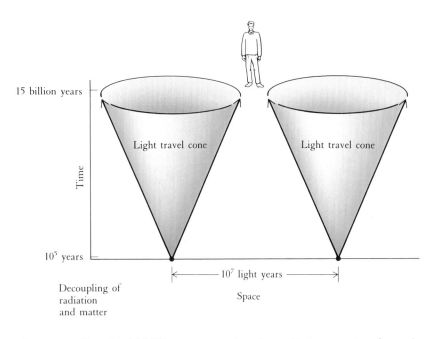

been traveling for 15 billion years, as has the radiation coming from the other direction. Thus, the two points that emitted that radiation are now approximately 30 billion light-years away from each other; yet the universe is only about 15 billion years old. (Detailed analysis shows that when the light was actually emitted, the regions were separated by almost 100 times the distance that light could have traveled.) How could those points have had such coherent properties of temperature and density, even though there is no way that a signal could have gone from one of those points to the other? The universe is thus remarkably smooth on scales beyond which communication was possible.

Traditional big-bang cosmology simply assumes that the universe is homogeneous and isotropic. But how did it get to be that way? Where did the initial conditions come from that yielded a homogeneous, isotropic universe? It would be very nice if we could come up with a model that would naturally evolve into such a situation without our having to merely postulate it.

THE LUMPINESS PROBLEM

Another cosmological question, which appears at first to contradict the smoothness problem, is: How did the universe get to be so lumpy? We saw in the smoothness discussion that on the very large scale the universe seems to be extraordinarily smooth, down to less than 1 part in

10,000. We know, however, that on the small scale the matter in the universe is clumped into stars, galaxies, and clusters of galaxies. To make these clumps, the smooth cosmological background material had to somehow segregate and gather together. The normal picture is that some small volumes that were slightly more dense than their surroundings eventually grew by gravity into galaxies, and fragmented and condensed further into stars and planets. Where did these initial fluctuations in the density come from? Until recently these fluctuations had been assumed to be just the way the universe began. Although the universe was almost homogeneous and isotropic, there must have been some source of minute fluctuations of density that would eventually form the galaxies.

The brightest central members of the Virgo cluster of 1,000 galaxies. Our galaxy and the Andromeda nebula are the principal members of the "local group," which is on the distant edge of the Virgo cluster 60 million light-years away.

THE AGE PROBLEM

Although there is some debate about the precise age of the universe, there is almost total agreement that it is somewhere between 10 and 20 billion years old. The standard big-bang model tells us that the density of the universe will evolve on a dynamical time scale, the period of time over which the force of gravity can cause significant effects. In particular, the universe's dynamical time scale is the amount of time needed for density to evolve a noticeable amount. At present, the density can evolve on a time scale of tens of billions of years (the present age of the universe). In fact, the dynamical time for the universe is approximately equal to the age of the universe at any given time.

Recall, from Chapter 5, the definition of the parameter Ω (omega) as the ratio of the density of the universe to the critical density. If Ω is greater than 1, the universe has sufficient gravity that expansion eventually stops and begins to recompress; meaning that we have a closed universe. In this case, Ω will eventually rise very rapidly as the density increases, and the collapsing universe goes to a big crunch. If Ω is less than 1, then Ω will fall, approaching 0 as the expansion continues to the future big chill. Only if Ω is near 1 will it remain constant and go neither to infinity nor to 0. After 15 billion years, we still do not know the answer, which means that Ω is currently close to 1; so it must have been extraordinarily close to 1 in the very earliest moments of the universe.

Suppose Ω were not precisely equal to 1. How fast would it increase to infinity or decrease to 0? It turns out that Ω's rate of movement is approximately equal to the dynamical time, that is, to the age of the universe. Right now, if Ω were different from 1, that is, if Ω were 0.5 or 2, it would take a few times 15 billion years for Ω to become infinite or 0. But when the universe was, say, only 10^{-43} seconds old, it would have taken only a few times 10^{-43} seconds for Ω to become either infinite or 0.

We can extrapolate our laws of physics back to a cosmological time of only 10^{-43} seconds. At that moment, space would have been so dense that black holes would form everywhere, and quantum physics would need to be applied to gravity. This instant is as far back as we can know anything about our universe using physics as we now understand it. It is the smallest unit of time as we know it.

Our newborn universe was evolving in its initial stages with a time scale of the order of 10^{-43} seconds. Thus, Ω should have been either increasing to infinity or approaching 0 on a time scale of 10^{-43} seconds, depending on whether or not Ω was greater than or less than 1. In either case, the lifetime of the universe could have been short indeed! Only if Ω equaled 1 would it be able to stay at 1 without evolving in one direction or the other. That is, $\Omega = 1$ is a very unstable dividing point.

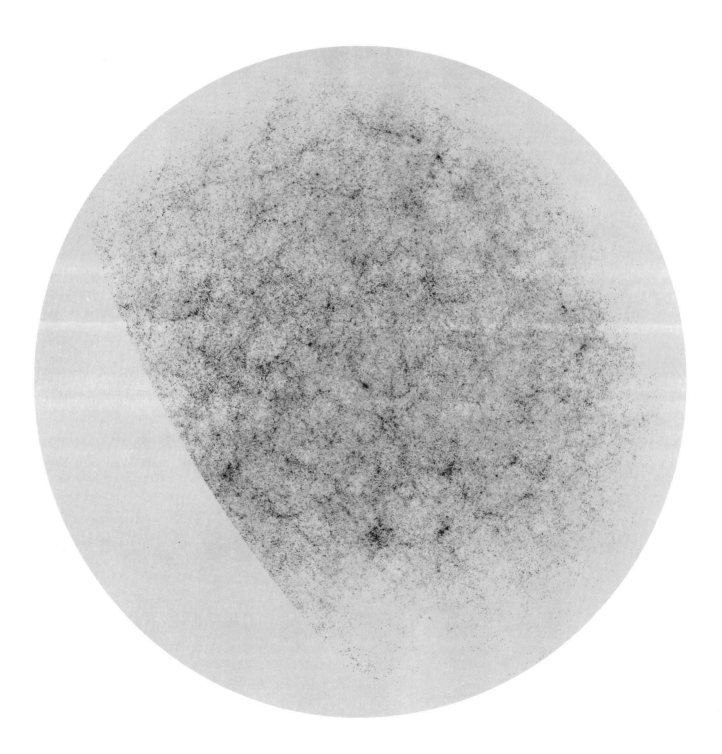

Note the appearance of structure even on this two-dimensional projection showing about a million galaxies, whose positions were observed at Lick observatory in California. The plotting routine for this projection was developed by P. J. B. Peebles of Princeton University.

The present age of the universe is about 15 billion years—much, much longer than 10^{-43} seconds. In fact, it's more than 10^{60} intervals of 10^{-43} seconds each. Hence, if the universe had had a value for Ω differing from 1 near the sixtieth decimal place, we would just now have Ω evolving away from 1. At present, from observations of galaxies and their dynamics, we don't know for sure if Ω is exactly 1, or if it is 0.1 or even as high as 2 or 3. On scales of 0 or infinity, these values are still remarkably close to 1. We do know that Ω must have been extremely close to 1 at the time when the dynamical time scale was very rapid (about 10^{-43} seconds). Somehow, the universe had its initial Ω very finely tuned to a value extraordinarily close to 1 (to an accuracy of about sixty or more decimal places).

Another way of stating the problem is that the expansive force tending to make Ω approach 0 and the contracting (gravitational) force tending to make Ω approach infinity must have been equal to an unbelievable precision in order to have the universe be still finite and feeling fine.

That the universe has lived 15 billion years without Ω equaling either 0 or infinity is sometimes referred to as the flatness problem, since $\Omega = 1$ describes a geometrically flat universe rather than one with curved space-time. Again, there ought to be some way to understand why Ω in the early universe is very close to 1, instead of requiring this to be an arbitrary starting condition.

THE ROTATION PROBLEM

The rotation problem questions why the universe is not rotating. In fact, to rotate you have to talk about what you are rotating around. But the relatively uniform Hubble expansion tells us that the universe has no center—no preferred point. All points are equivalent. How did it get to this state, with no preferred point, with all points being equivalent, with no axis of rotation?

THE MONOPOLE PROBLEM

Here's one more puzzle that seemed initially to have no solution. Although our GUTs did a fine job of explaining the excess quarks, they have one nasty side effect: They also make a particle called the magnetic monopole. A magnetic monopole is a pure magnetic charge. Although we see pure electric charges on protons and electrons, we never see a pure magnetic charge. One never sees a magnetic north pole without a magnetic south pole in tandem. All GUTs, when operating in the context of the standard big bang, yield almost as many of these monopoles as there are neutrons and protons. The monopoles have masses of about 10^{17} GeV and have magnetic charge. They were made in the high-energy period of the early universe and should still be around. There have been many experimental searches for these monopoles, but they have not been found. Certainly there cannot be as many of them as there are neutrons and protons; so what is wrong with the GUTs?

THE SOLUTION: INFLATION

The aforementioned problems of smoothness, lumpiness, age, rotation, and monopoles refer to some early initial conditions out of which the universe emerged. Past attempts at solutions have all required some very special arbitrary setting of those conditions. The "inflation" solution to all of these problems was proposed in 1980 by Alan Guth, now at M.I.T. As mentioned before, inflation refers to a phase in the evolution of the universe when its rate of expansion was extremely rapid, doubling each 10^{-34} seconds. How this comes about is another example of how the inner-space/outer-space connection works. Inflation is not an alternative to the big bang. Instead, it enables a wide variety of arbitrary initial conditions to converge to the special initial conditions that we observe for our universe. It does this convergence at times of about 10^{-34} seconds during the GUT epoch.

We will first describe the inflation phenomenon and then show how it solves many of our problems.

To understand how we can have inflation, let us note that Einstein's general theory of relativity tells us that the rate of the universe's expansion is directly related to the density of matter and radiation in it, and since mass and energy are equivalent ($E = mc^2$), this means that the universe's expansion is related to energy density: the higher the universe's energy density, the higher its rate of expansion; and conversely, the lower the energy density, the lower the rate of expansion. Because the early universe was very dense, it initially expanded very rapidly. (It had to or it wouldn't have expanded at all). In the standard model for cosmological expansion, as time went on, the density decreased, and so the rate of expansion decreased. The reason that the density decreases is that expansion adds space, so the mass-energy is spread out over a larger and larger volume of space. The decrease in the energy density decreases the rate of expansion yet further. But the universe could inflate at a constant rate if somehow the density remained constant, so that even though the distance scale gets bigger, the amount of energy per unit volume remains the same.

How is this possible? If the energy density is made out of matter and radiation, adding volume without adding matter and radiation would decrease the energy density. However, what would happen if the space that is being added contained huge amounts of energy density? That is, what if the space being added to the universe had some finite value of mass-energy density, rather than a zero density? The result would be that the space that is being added is also adding energy.

Just such a mechanism, an energy-containing space, was discussed in Chapter 4, in the form of a Higgs field. Recall that this field was proposed as nature's mechanism for hiding gauge symmetry and giving some of the force-carrying gauge particles a mass. The universe begins in a state without the cosmic Higgs-like field. As it cools, it passes below a critical temperature and the Higgs-like field appears. It is near this transition that Higgs-like gives energy to the vacuum. This is just what is needed for cosmic inflation. Cosmologists have a special term for this kind of vacuum: They call it a *false vacuum*. (Unlike the classical vacuum, this one has energy density, but in all other ways it is similar.) The Higgs field we met earlier concerned the electroweak forces. Now we have encountered a similar kind of beast which is crucial to cosmic inflation. Finally we note that with the inclusion of the strong force, grand unification requires a Higgs mechanism.

The rate of expansion of the universe has several phases, depending on how much energy is brought in by the added space. In the earliest phase (hottest temperatures, time less than 10^{-35} seconds), the radiant energy density is so enormous that any added vacuum has little effect on the rate of expansion. During this early time the expansion of the universe decreases its energy density in the familiar way.

At the "cooler" GUT level, the cooling universe passes a crucial temperature and the vacuum energy becomes significant compared with the radiation density, and we are in the inflationary phase. Inflation ends when the Higgs-like field becomes nonzero and breaks the symmetry. This differentiates the strong force from the electroweak force (see the figure on the following page). Thus, the phase change that arrests inflation changes false vacuum to *true vacuum*, where space has no intrinsic energy density (see the figure on page 175), and changes symmetry to broken symmetry. The expansion slows to that of the original big-bang model. The enormous energy contained in the false vacuum now generates particles. A dense "soup" of elementary particles forms, expands, and cools, and the standard-model big-bang theory begins.

Now, how does the inflation phase solve the many "initial condition" problems we listed earlier?

The vacuum-driven expansion will result in a tremendous increase in the size of the universe in a very short time. In fact, inflation during

The vacuum energy density of the universe is like an "effective potential energy" for the universe, and all systems want to minimize their potential energy. These energy density curves represent the "effective potential energy" of the system for different symmetries, or unification of forces. At high temperatures, the unified forces of nature produce perfect symmetry at the minimum energy state. The universe has a positive energy density in this symmetric state. At lower temperatures, the minimum energy occurs when symmetry has been broken. The energy density of the minimum drops to zero, indicating a vacuum like ours with no energy density. The false vacuum with positive energy density is what drives inflation.

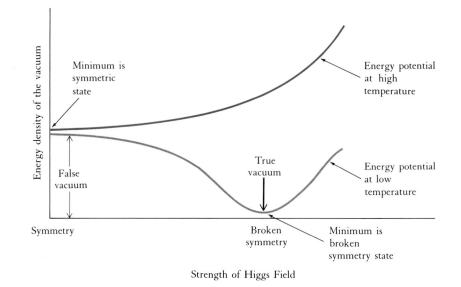

this epoch would probably have increased the size by a factor of over 10^{50}. Inflation starts when the universe is about 10^{-35} seconds old, and more than doubles the universe in size every 10^{-35} seconds until about 10^{-32} seconds; that is, the universe's size is doubled more than 1,000 times, so the total increase is more than 10^{50}. This much expansion can solve the previously mentioned initial-condition problems. In particular, the smoothness problem is solved because the region that we are starting with is now very small within which everything is in causal connection. (The radius of the universe is 10^{-35} light-seconds at the start of the GUT's epoch, when the vacuum energy would be dominant.)

During inflation, that same small region becomes so large that all the matter in the present universe comes from that single little region of 10^{-35} light-seconds. Since all the regions of the presently known universe were causally connected at that early time, we can thus understand how they can all be at the same temperature today. The age problem is satisfied because inflation makes things so big that any local region that may have originally been curved looks flat, and a flat universe has Ω equal to 1. This is analogous to why the Earth appears so flat to a person standing in a field looking at the horizon (especially in Illinois), even though on a large scale it is round. The rotation problem is also satisfied, because the terms involving rotation about any axis are also proportional to the curvature, and they disappear as the local curvature approaches 0 (flat).

The monopole problem is solved because one monopole per horizon volume is produced just before inflation. A *horizon volume* is the volume of the universe that is causally connected; its radius is the dis-

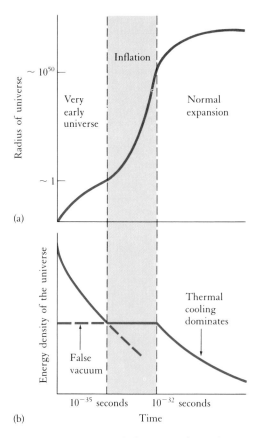

(a) The separation of objects in the universe increases at a much greater rate during the inflationary period. (b) The density of the universe steadily decreases, except during the inflationary phase when it is dominated by the false vacuum. Note that when the density stays constant and does not decrease with expansion, this causes the universe to inflate.

tance that light can travel in the available time. Just before inflation, the horizon volume is a sphere of a radius of about 10^{-35} light-seconds. If we multiply the horizon volume by the density at that time, we find that the horizon volume then contained only about 10^{-80} of the matter in today's horizon volume. Thus, in the standard model, we had to add many (approximately 10^{80}) of those horizon volumes together to get our present epoch. We would then end up with about 10^{80} monopoles, a number roughly equal to the number of protons. However, in the inflating model, we make our universe out of only one pre-GUT horizon volume. Thus, there would be only one monopole in the entire present universe, and we could never expect to find it. (Actually, there may be a few more because of collisions and events that might occur slightly after inflation, but there would be very many fewer than 10^{80}.)

It is curious that it was the monopole problem that Alan Guth originally set out to solve. But in finding a solution to it with his inflationary mechanism, he simultaneously found a solution to the long-standing problems of the universe's initial condition: smoothness, age, and rotation. In fact, Guth, a particle theorist, had only heard of the age–flatness problem a few weeks prior to his coming up with the solution. This illustrates the importance of interdisciplinary communication in academic fields. New points of view can get one around apparently insurmountable walls.

You might now ask, If the monopoles are "inflated away," what about baryons? since baryons were produced at the GUT epoch as well. However, the baryons were produced just as the GUT epoch ended, when the universe's baryon nonconservation forces ceased being in equilibrium. The baryons were produced at the end of inflation, rather than at the beginning, so all of the baryons in the universe could be made without any problem. (It is important to note that until we knew when and how the baryons were produced, an inflationary model was impossible, since it would have predicted an empty universe.)

Inflation goes a long way to making the universe "natural," that is, a result of the laws of nature, rather than the product of an incredible series of miraculous starting conditions. There are, however, a few remaining problems. One, mentioned earlier, is the lumpiness problem. If we smooth things out so much that we don't have any lumps left to make galaxies, we have a difficulty. Another problem is: Once inflation is going, how is it stopped? Once expanding at this extraordinarily rapid rate, how does it ever go through a phase transition? As we mentioned before, in the symmetric phase of the universe all the forces were equivalent. In its broken symmetric phase, the forces were different. The passage from symmetry to broken symmetry is the *phase transition*. How do we get from one to the other? We know that as the universe cooled, symmetry was broken. To have the phase transition go to *completion*, each region of space where the symmetry has been broken must unite with the other regions of space where the symmetry has been broken. It is as if you were trying to convert a pool of water into solid ice (another example of a phase transition). The little crystals of ice in the water gradually merge together until the whole pool becomes ice. What would happen if the water was flying apart and expanding faster than the ice crystals were growing? Then you would have a snow-making machine and could not produce a solid block of ice. The ice crystals, the granules of the new phase, would never merge together.

People who work on phase transitions refer to these granules of the new phase as *bubbles*. For a phase transition to go to completion, the bubbles of the new phase have to all run together. The bubble analogy comes from steam forming out of water. With steam, water goes to a higher-temperature phase, and the bubbles of water run together, converting all the water into steam. Whenever there is a phase transition in our normal space, the bubbles, or grains, of the new phase all eventually merge together. With cosmological inflation, this might not occur, because the bubbles of the new phase would be moving apart from each other faster than they are growing. Thus, we might never get a GUT phase transition going to completion.

If there were a phase transition going to completion, we would have a way of understanding the origin of the universe's lumpiness, that is, galaxies. They would come from the merging of the different granules, or bubbles. This would be analogous to the little white lines of imperfections in ice cubes that result when the different freezing grains met but did not quite match up. Cosmologically, such imperfections could be the "lumps" that eventually grow into galaxies and other astronomical objects. However, in the original inflationary model, it appeared that the bubbles of the new broken-symmetry phase moved apart from each other faster than they could run into each other; so the transition never went to completion, and no imperfections of this kind could occur.

THE NEW INFLATION: BUBBLE UNIVERSES

A solution to this difficulty was proposed independently by Andre Linde, in the Soviet Union, and Paul Steinhardt and his student, Andreas Albrecht, at the University of Pennsylvania. Their proposal is that the cosmological phase transition is of a special type first proposed by Erik Weinberg of Columbia University and Sidney Coleman of Harvard. Whereas conventional phase changes are abrupt, the type proposed here is smooth and slow. The type of phase change is basically determined by the choice of parameters of the grand unified theory. As long as our experimental knowledge here is incomplete, we can choose these parameters as we see fit to produce a satisfactory theory of the evolution of the universe. (We would in turn argue that the universe also "chose" those parameter values.) A smooth phase transition would enable the entire universe to be one bubble; there would not be separate bubbles that have to catch up with each other. By having the phase transition be particularly smooth, the energy density would not be concentrated on the walls of the bubble, as in a sharp phase transition, but instead would be uniformly spread throughout the bubble. The new bubble would appear to be homogeneous and isotropic. The smooth phase transition would also enable enough excess quarks to be produced within the new bubble to be consistent with what exists in the present universe.

The concept that the universe is a single bubble has tremendous implications: There may be many other bubbles out there, all of which could be other universes, completely disconnected from ours. This type of solution involving multiple bubbles, with our universe being within a single bubble, was also proposed by Richard Gott of Princeton on geometric grounds. He argued that a single bubble solution seemed to be most consistent with the geometric requirements of an early universe.

There may be more universes than we ever contemplated, but we will have no way of reaching them. Even "Star Trek," with its warp drives that move from galaxy to galaxy at amazing, nonphysical, faster-than-light velocities, would not be able to use such concepts to get from universe to universe. The type of space between bubble universes is not normal physical space at all, but a space where all the forces are unified; a region where protons decay instantaneously, quarks and leptons interchange freely, and normal matter does not exist. There is only energy, and that energy has no memory, in the sense that it is, in fact, all states of matter at once.

These new inflationary bubbles appear to have one problem— lumpiness. If the entire universe is one bubble, then that one bubble would have to create all the lumps. We could not create them by collisions of bubbles. And since it would be one, single inflating bubble, we could not rely on any lumps that were produced earlier, say, during the epoch when gravity was unified with the GUT force in a theory of everything (about which we understand even less than we do for the GUT epoch). No lumps from this earlier epoch would survive past the inflation. They would be inflated away and smoothed out in the normal inflating manner. The perturbations that would eventually grow into clusters of galaxies would have to be created within the single bubble.

In 1983, several research groups around the world (including Guth, Steinhardt, and Linde, as well as Stephen Hawking at Cambridge and Michael Turner at Chicago) began to look at this problem. Interesting proposals came forward. It was found that in this single bubble, there

A computer simulation of mass points assembling by gravity to create structure. The comparison of such simulations with the actual structure of many galaxies and clusters is a method for evaluating the validity of theoretical ideas about galaxy formation. This simulation was carried out by Adrian Melott of the University of Kansas.

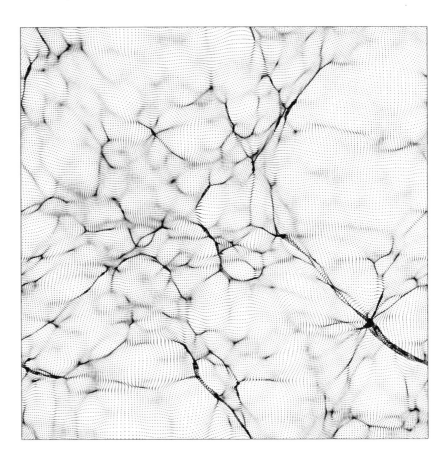

would be variations in the energy density caused by random quantum fluctuations at the GUT epoch, when the matter that makes up today's universe was compressed into a region smaller than a present-day soft-ball—a region so small that on the scale of galaxies and clusters of galaxies, quantum effects and quantum uncertainties would be important. The quantum lumps that would spontaneously form, in fact, were found to have a very nice feature. They had exactly the mass spectrum predicted years earlier by the Soviet cosmologist Yakov Zel'dovich and the English cosmologist Ted Harrison. Harrison and Zel'dovich hypothesized a primordial distribution of lumps with a relative grouping of sizes that approximates the present-day arrangement of galaxy masses.

This theoretical explanation was, however, not without its problems. Although the quantum lumps had the right relative ratio of masses, they produced such large fluctuations that the lumps immediately formed black holes rather than slowly growing into galaxies.

A three-dimensional "slice" of the sky observed by Margaret Geller, John Huchra, Valerie de Lapparent, and R. McMahan of the Harvard–Smithsonian Center for Astrophysics. The points are galaxies. Note the structure, including clumps and voids. These structures are at least 30 million light-years across.

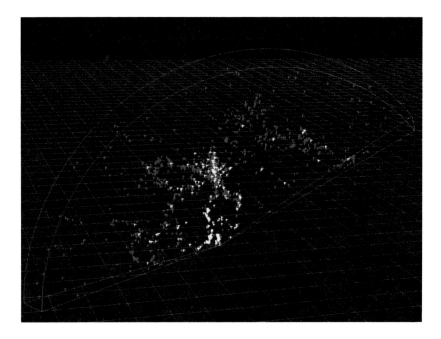

At present no satisfactory theory easily gives the correct size lumps at the end of inflation without special fine tuning. Some have argued that instead of random lumps, GUTs might produce energy concentrations in long stringlike configurations. These three-dimensional cosmic strings are remnants of the GUT phase transition. In fact, they are little tubes of the high-energy vacuum, which, in some theories, can survive to the present epoch while embedded in a normal, zero-energy vacuum. Loops formed from these tubes gravitate and thus could become seeds to trigger the formation of galaxies and clusters. Some have even speculated that these seeds can explode and push matter about to form galaxies and clusters. Another option is that some other late phase transition might superimpose structure on the matter.

In any case it seems clear that astronomical observations of the large-scale distributions of galaxies and clusters of galaxies might tell us much about the origin of lumpiness, which in turn might tell us about early-universe phase transitions. Margaret Geller and John Huchra and their colleagues at the Harvard-Smithsonian Center for Astrophysics in Cambridge, Massachusetts, have been making painstakingly precise three-dimensional maps of slices of the sky and coming up with remarkable threadlike distributions of galaxies surrounding huge voids. In this way, telescopes are beginning to tell us things about fundamental physics and augment the information gleaned from accelerators. Already telescopes are telling us that there are extraordinarily large correlated structures in space, structures that are increasingly difficult to explain in terms of random clumping.

INFLATION CRIES OUT FOR SUPERSYMMETRY!

Cosmologists and particle physicists were not stopped by the problems of making the structures they observed compatible with inflation. They immediately looked for grand unified models that would not produce black holes, but would still have all the appropriate properties of the single-bubble universes and still inflate sufficiently. This set of requirements became an interesting way to select between the different possible grand unified models.

The current leaders among such models are the so-called supersymmetric (SUSY) grand unified theories (GUTs) or SUSY-GUTs models. SUSY models seem able to keep the quantum fluctuation amplitudes small while doing everything else appropriately. This was first pointed out by the CERN theorists, John Ellis, Keith Olive, Dimitri Nanopoulos, and Mark Srednicki, in their paper entitled "Inflation Cries Out for Supersymmetry!" As we mentioned before, supersymmetric GUTs models had been proposed much earlier, as a possible way to bring gravity into grand unification.

In SUSY models there is a pairing of fermions (quarks and leptons) and bosons. The proposed supersymmetric partners for the currently known fermions are called *squarks* and *sleptons*. The proposed fermionic partners for the bosons have names like gravitino, photino, Wino, and Zino, and for the GUT's gauge bosons, Xino and Yino. Of course, for the Higgs field, we must have a Higgsino. Thus, SUSY predicts a veritable zoo of new particles. These new particles give enough freedom to the grand unified theory to minimize the size of the quantum lumps that developed, but still have all the properties of inflation, and satisfy all the other conditions. This kind of model also proposes possible new candidate particles for the dark matter of the universe. In Chapter 7 we will discuss the problem of detecting these hypothetical SUSY particles as well as other dark-matter candidates.

Not only is inflation able to solve the cosmological initial-condition problems, it makes one dramatic prediction. Recall that $\Omega = 1$ is the boundary between an ever-expanding universe ($\Omega < 1$) and an eventually contracting universe ($\Omega > 1$). When inflation solves the age (flatness) problem, it makes Ω exactly equal to 1 out to a hundred or more decimal places! Inflation predicts the value of Ω, as no other physical theory has been able to do. In fact, past observational evidence did not enable astronomers to measure Ω precisely. To some extent they had to guess at its value. Inflation gives a model that specifically predicts Ω to be exactly equal to 1 to an accuracy better than any conceivable observation. When we looked at big-bang nucleosynthesis, we noted that the light-element abundances fit observations only if the amount of matter that enters into nuclear reactions is about 10% of the critical value. All baryonic matter enters into nuclear reactions; so big-bang nucleosynthesis tells us that the value of Ω for baryons is approximately 0.1. The abundances of elements like deuterium are not consistent with observations if this value for Ω is near 1. If inflation is right, and the total value of Ω is equal to 1, then the bulk of the matter in the universe would be nonbaryonic (perhaps it would be made of SUSY particles).

PROTON DECAY: UNDERGROUND PHYSICS

Another spin-off of supersymmetric grand unified models is their prediction of proton decay into the most massive quarks allowed by the mass of the proton. This would mean that proton decay products would include the strange quark, rather than up and down quarks. The "traditional" proton decay would be into an antielectron (positron), and a pion, which is made of up and down quarks:

$$p^+ \rightarrow e^+ + \pi^0$$

Grand unified theories would allow protons to decay. In one proposed decay route, shown to the right, the proton's constituent u quarks combine to form an X particle, which disintegrates into a d antiquark and a positron (a lepton). The d antiquark combines with the remaining quark of the proton (a d quark) to form a neutral pion. Pions decay, and positrons are soon annihilated in matter–antimatter collisions, releasing energy.

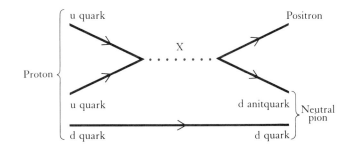

A physicist is checking the photomultiplier tubes in the Kamioka proton decay detector prior to the addition of water. The detector is deep underground in a zinc mine, where it is shielded by the Earth from most cosmic rays. These photomultiplier tubes detected the flashes of light produced by a burst of 11 neutrinos hitting the water in the detector after traveling 170,000 light-years from Supernova 1987a.

With supersymmetry the main mode would create antimuons, and kaons, which have strange quarks:

$$p^+ \rightarrow \mu^+ + K^0$$

Although various proton-decay experiments have not seen any positron–pion modes, the limits on muon–kaon modes are not as stringent.

The only laboratory experiments to date that probe the realm of GUTs and SUSY are not accelerator experiments. Instead, they consist of large detectors placed deep underground. These underground detectors were designed to search for the decay of protons, as we noted in Chapter 5. However, the proton lifetime predicted by even the most optimistic GUT theories is greater than 10^{30} years (far longer than the age of the universe). Obviously, we cannot wait 10^{30} years to see if a proton decays. What we can do is build a detector that contains 10^{33} protons and see if any of them decays during a year or so. Because proton decays are so rare, it is important to insulate the experiment from any other possible reaction that might take place in the detector. Hence, the detectors were placed deep underground to shield them from the ever-present cosmic-ray interactions.

There have been proton-decay detectors built in deep mines and in tunnels under mountains. For example, in addition to the Kamioka Zinc

A large new underground laboratory is being created out of solid rock off the Grand Sasso Automobile Tunnel in central Italy.

Mine in Japan and the Morton Salt Mine near Cleveland, which we discussed in the first chapter, there are detectors in the Kolar Gold Mine in India, and the Soudan Iron Mine in northern Minnesota. There are also detectors in alcoves of the automobile tunnels under Mt. Blanc, and near Frejus at the French–Italian border, and a new detector is being installed in the auto tunnel under the Gran Sasso in Central Italy. The largest and currently most sensitive of the operating detectors are the Morton Salt Mine detector, operated by a collaboration of scientists from the University of California at Irvine, the University of Michigan, and Brookhaven National Laboratory; and the Kamioka Mine detector operated by a collaboration of scientists from the University of Tokyo and the University of Pennsylvania. Each of these detectors consists of a large volume of water (about 10,000 tons, which contains about 10^{33} protons) rigged with a large array of photomultiplier tubes, which can detect tiny flashes of light in the water.

When a proton decays to a positron and a pion it produces a flash of light because of the subsequent decay of the neutral pion into two gamma rays. The gamma rays convert in the water to electron–positron pairs. All of the fast-moving charged particles emit Cerenkov light which reaches the photomultiplier tubes. Radiation from the positron adds to the signal. The primary source of background radiation is neutrinos generated in the Earth's atmosphere by cosmic-ray protons. These can penetrate any depth of rock, and can occasionally cause collisions in the water detector. The rates of such collisions would roughly correspond to the same rate at which proton decay would be seen if protons lived about 10^{31} years. To distinguish these events, we can note that the total energy of a proton decay would be 1 GeV, that is, the rest-mass energy of the proton; whereas neutrino collisions yield a wide distribution of energies. Also, the total momentum in proton decay is zero since the proton is sitting still in the water, whereas neutrino collisions would bring in momentum. Thus, the designers of the detectors had to plan to gather much detailed information on the events. This is made difficult by the huge volume of the detector, every point of which could be the site of a possible event to be examined in detail. The detectors working in 1988 used arrays of hundreds of large photomultipliers, which provided at least limited spatial detail on all the atoms in the detector.

None of these detectors have seen signals consistent with proton decays even after several years of operation. The combined result teaches us that the mean life of the proton in water to this mode of decay is longer than 10^{32} years. However, this type of detector is not as sensitive to other modes of proton decay; so the muon–kaon mode preferred under supersymmetry may still be possible. To attempt to get better data on muon–kaon modes, new detectors in the Frejus and Gran Sasso tunnels and in the Soudan mine are being developed; these use media other than water, and gather much more detail on the particles emerging from a proton decay. We saw at the very beginning of this book how underground detectors designed to test GUTs succeeded in detecting neutrinos from Supernova 1987A. However, the jury is still out on grand unified theories, which predict the instability of the proton.

GRAVITY ENTERS: THE THEORY OF EVERYTHING

We noted before that at sufficiently high temperatures, we expect that the gravitational force will be unified with the other forces. At this epoch, the universe was of subatomic dimensions, and so only a quantum theory of gravity is relevant. Compared with how other forces are quantized, we expect this event to have occurred when the temperature of the universe was about 10^{19} GeV (10^{32} K). The densities then would have been around 10^{92} grams/cubic centimeter. At these enormous densities, every point in the universe would have instantaneously become a black

At the Planck time, all of space-time is a sea of exploding and reforming black holes. Different parts of space-time become connected and separated from each other spontaneously.

hole, but these little black holes would then have evaporated (see figure on this page). Their rate of evaporation would have been the same as the age of the universe at that time, approximately 10^{-43} seconds, which we call the Planck time. The entire universe would have been continually in the state of a space-time foam, forming black holes and exploding them. Since each black hole connects and disconnects regions of space-time from each other, the explosions and formations would leave all of space-time very convoluted. Space-time itself would have become quantized and no longer continuous. This is very different from our normal way of looking at things.

At present, we do not know how to do physics in such an epoch, and the known laws of physics are not understandable in the space-time foam. It appears that our extrapolation to high temperatures and densities runs into difficulty here. In fact, even calling this time (10^{-43} seconds) the age of the universe is really quite misleading, since we cannot extrapolate time to any earlier epoch. It may be that time as we know it did not exist prior to this quantum gravity epoch. Whatever we mean by *prior* loses meaning, since *prior* is a time-linked word. The standard question, "What was there before the big bang?" may be a cheat, since *before* implies that we have some concept of time in these extraordinary conditions.

Probably there was some sort of phase transition as the universe went from this quantum gravity epoch, where all of the forces including gravity were unified, to the broken symmetry where gravity separated out, at around 10^{19} GeV. We don't hesitate to admit that we don't understand these murky beginnings, but it is nice that our lack of understanding does not affect the things that happened later. Space-time as we know it was clearly applicable back to some early time. As long as known lower-temperature interactions were in equilibrium at that time, with both forward and reverse reactions occurring, we lose all trace of any prior initial conditions, so that we only need to go back to the time of such equilibrium to understand later events. This is how we estimate the number of neutrons in big-bang nucleosynthesis. All we had to know was that the neutrons and protons reached an equilibrium requiring a temperature around 10^{10} degrees Kelvin. Similarly, for the origin of matter, all we need is an equilibrium of the quark-producing interactions. That required temperatures of approximately 10^{14} GeV (10^{27} K). Although this is a high temperature, we still do not need knowledge back to about 10^{19} GeV (10^{32} K), when the Planck-time uncertainties would have entered.

STRINGS AND SUPERSTRINGS

One of the most promising aspects of quantum gravity theory in recent years has been the concept that perhaps the fundamental entities of space-time are not points but small loops of "string": in other words, one-dimensional (hence string) rather than zero-dimensional points. In particular, John Schwarz of Cal. Tech. and Michael Green of London showed that supersymmetry theories could eliminate certain mathematical problems if the fundamental entities were ten-dimensional "superstrings" instead of three spatial points moving in a fourth dimension we call time. Edward Witten, of Princeton, argued that such a theory was the only possible complete supergrand unified theory, and many have said that superstrings are the ultimate theory of everything (TOE). Of course, even superstrings have some problems—such as getting from the ten-dimensional superstring space to only three space dimensions and one time dimension. However, scientists are more hopeful that some sort of TOE might eventually be found now that superstring theories exist.

These are the theoretical issues that have converged inner and outer space. In Chapter 7 we will consider the instruments and experiments now being planned not only to test these theories but to attempt to solve the puzzles and paradoxes that currently challenge us.

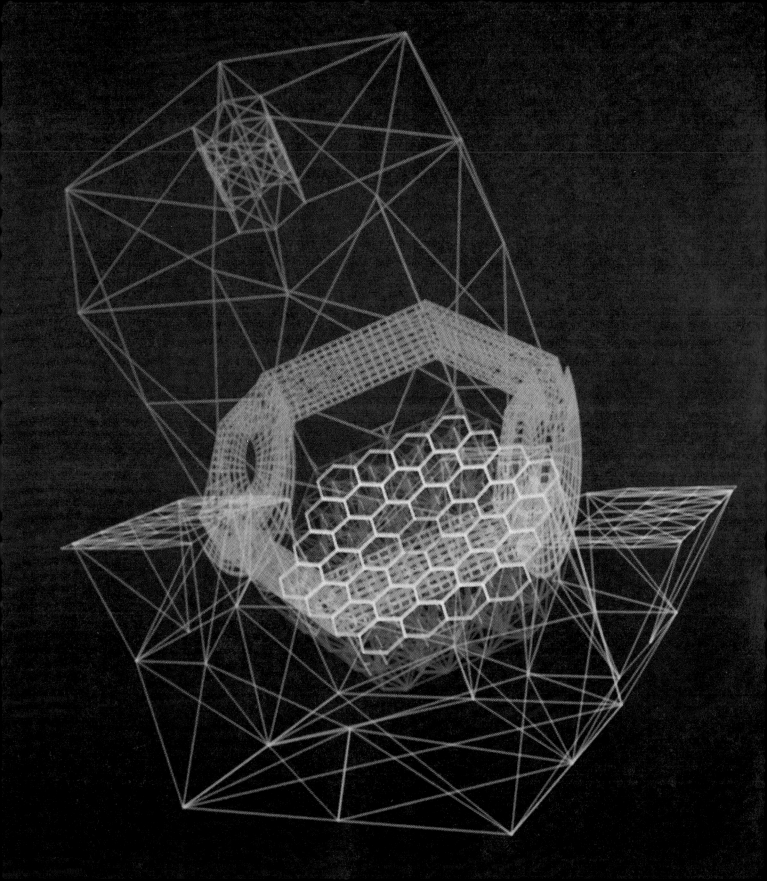

7 TOOLS FOR THE 1990s

An engineering drawing of the Keck telescope now under construction in Hawaii by Cal. Tech. and the University of California. The 10-meter-diameter mirror will be composed of 36 hexagonal segments.

In this chapter we will discuss how detectors, accelerators, telescopes, and ancillary devices will shape the course of particle and cosmological research until the end of this century.

In Chapter 4 we described the standard model (SM) of particle physics. In all, some twenty numbers (particle masses, strengths of the forces, etc.) are needed to rigorously describe a universe run by the standard model. These numbers are often poorly known. Finding precise values for them will require either new accelerators or greatly modified existing machines. In Chapter 6 we sketched many proposed extensions of the standard model, each of which predicts the existence of new particles of larger and larger mass. For example, the Higgs particle may exist at the very high mass of 1 TeV, that is, 1,000 times heavier than the proton. Finding such problematic particles requires very sophisticated accelerators. We can have several different designs for accelerators, but each has its own problems and advantages. We could also have several different strategies for probing particles: We could have protons hitting a fixed target, protons colliding with protons or with antiprotons, electrons colliding with positrons, and, finally, electrons colliding with protons. A final factor in choosing an accelerator is the issue of energy versus luminosity. Luminosity is a measure of the number of collisions taking place per second.

Let's look at the virtues and the shortcomings of some of these approaches. Since modern accelerators cost hundreds of millions, or even billions, of scarce dollars, and thousands of even scarcer man-years of creative scientific and engineering effort, the options must be considered carefully.

PROTON (HADRON) MACHINES VERSUS ELECTRON–POSITRON COLLIDERS

Physicists now believe that the electron is a "point" object, with no internal structure. Electron–positron collisions are "clean," in that only the fundamental particles themselves are interacting. Furthermore, the energy in the collision is the sum of the energy in each beam. For example, the Positron–Electron Project (PEP) machine at Stanford collides 15-GeV positrons against 15-GeV electrons, and studies particles and phenomena at a total collision energy of 30 GeV. That's good.

A proton is a bag of quarks and gluons, each of which shares some of the energy given to the proton. The most interesting collision between two protons occurs when a quark (or gluon) in one collides with a quark or gluon in the other. The remaining quarks and gluons act as "spectators," absorbing relatively little energy from the collision, but still spraying out and burdening the detectors with tracks. This yields very "dirty" collisions that are hard to interpret. That's bad.

Also, the objects participating in the collision have only a fraction of the proton's energy, so that, in general, the total collision energy is much less than the sum of the energies of the protons. On the average, detailed calculations show that the loss in collision energy due to the complex structure of the protons is about a factor of 6 (approximately the number of constituent quarks in the incoming particles). At the Fermilab Collider, the fundamental collisions *average* about 300 GeV, even though the colliding proton and antiproton energy totals 1,800 GeV. That's bad.

The redeeming feature is that proton acceleration is easier and cheaper than electron acceleration, so that proton energy of near 1,000 GeV has been achieved, whereas the highest-energy electrons ever accelerated are only 50 GeV. Thus, proton collisions are more complex and require more sophisticated detectors, but can still reach higher energy than electrons.

The forces between electrons are purely electromagnetic at moderate energies, but as the colliding energy approaches 90 GeV (mass of the Z^0), the weak aspect of the (unified) electroweak force becomes important. In contrast, in the collision of hadrons, such as protons, the dominant force is the strong force, the electroweak playing a significant but subtler role. Hence, neither type of collision is absolutely superior to the other in research terms. Which is better depends on what research problem you are trying to solve. In the table on the facing page we list the world inventory of particle physics machines; these indicate the decisions made in different regions.

ACCELERATOR INVENTORY

Name and location	Particle energies	Colliding particles
I. "Old timer" machines (but still operating for good physics)		
The National Laboratory of Japan (KEK), Japan	12 GeV	proton—fixed target
Alternating Gradient Synchrotron (AGS), Brookhaven, N.Y., USA	30 GeV	proton—fixed target (plus heavy ions)
Serpuhkov Proton Accelerator, USSR	70 GeV	proton—fixed target
Superproton Synchrotron (SpS), CERN, Switzerland	450 GeV	proton—fixed target
Cornell Electron Storage Rings (CESR), Cornell, N.Y., USA	8×8 GeV	electron–positron
Positron–Electron Project (PEP), Stanford, Calif., USA	15×15 GeV	electron–positron
VEPP IV, USSR	6×6 GeV	electron–positron
II. Brand new or nearly working machines		
Tevatron Fixed Target, Fermilab, Ill., USA	800 GeV	proton—fixed target (start-up: 1984)
Super $\bar{p}p$ Synchrotron (S$\bar{p}p$S), CERN, Switzerland	315×315 GeV	proton–antiproton (start-up: 1981)
Tevatron Collider, Fermilab, Ill., USA	900×900 GeV	proton–antiproton (start-up: 1987)
TRISTAN, Japan	30×30 GeV	electron–positron (start-up: 1987)
Stanford Linear Collider (SLC), Stanford, Calif., USA	50×50 GeV	electron–positron (start-up: 1989)
Large Electron–Positron accelerator (LEP I), CERN, Switzerland	50×50 GeV	electron–positron (start-up: 1989)
III. Under construction		
HERA, W. Germany	30 GeV $\times$ 1,000 Gev	electron–proton (start-up: 1990)
UNK I, USSR	3 TeV	proton—fixed target (start-up: ~1993)
Beijing Electron Storage Rings (BESR), People's Republic of China	3×3 GeV	electron–positron (start-up: 1989)
IV. Proposed, designed, or dreamed		
UNK II, USSR	3×3 TeV	proton–antiproton
LEP II, CERN, Switzerland	100×100 GeV	electron–positron
Large Hadron Collider (LHC), CERN, Switzerland	$6-8 \times 6-8$ TeV	proton–antiproton
Superconducting Supercollider (SSC), Tex., USA	20×20 TeV	proton–antiproton
ELOISATRON, Italy	100×100 TeV	proton–antiproton
Serpukhov Linear Collider, USSR	1 Tev $\times$ 1 TeV	electron–positron

FIXED-TARGET COLLISIONS VERSUS COLLIDERS

In an electron machine the fixed-target mode, popular in the 1960s, has been all but supplanted at the high energies by colliders. However, in proton machines it is useful to extract a beam of particles from the machine, transport it to a target, and use the ensuing collisions. As noted in the table on the previous page, as of 1988 there were five such facilities in the world and one under construction. The disadvantage of such a system is the large loss of available energy, not only because of the complexity of the proton, but also because the laws of conservation of energy and momentum require that the particles issuing from the collision carry away momentum equal to that brought in by the bombarding particle. Recall that when the arithmetic is done, a Fermilab 1,000-GeV proton, hitting a fixed target, cannot produce a new particle of mass more than 42 GeV. This large loss of effective energy is compensated for by the variety of research that can be carried out in fixed-target proton machines. Also, since all the protons can be made to interact in a dense target, the number of collisions is enormous, reaching as high as 10^{13} particles per second at the Brookhaven Alternating Gradient Synchrotron (AGS).

Let's see how a "typical" fixed-target program looks. At Fermilab, a large number of stable or relatively long-lived particles, produced in the primary collision, can be formed into secondary beams for subsequent experiments: pions, kaons, neutrons, and antiprotons are some of the popular secondary beams. The pions and kaons will decay in flight if the path length is many meters, and the decay products, muons, neutrinos, and photons, are formed into tertiary beams for different experiments. Finally, the targets can be varied from hydrogen to uranium, adding even more variety. To illustrate the variety of experiments possible using fixed targets, the table on the facing page lists experiments carried out during 1988–1989 at Fermilab. Another virtue of fixed-target operation is that more than fifteen experiments can be run simultaneously. Typically each of these experiments involves some fifty scientists and students and takes about 5 years from conception to completion.

Of course the colliders have a much more limited choice of colliding objects: protons versus protons or antiprotons, or electrons versus positrons. The great advantage is the much higher energy available for the exploration of new physics made available by the collider.

A good example of a fixed-target program that is closely connected to cosmology is one that involves one of the more exotic of leptons, the neutrino.

CURRENTLY APPROVED FERMILAB EXPERIMENTS FOR 1989

Experiment number	Scientific spokesman	Object of experiment
		FIXED TARGET

Electroweak forces

E-665	Montgomery, Fermilab	Muon scattering with hadron detection (13/79)*
E-782	Kitagaki, Sendai, Japan	Muon scattering with Tohoku bubble chamber (7/33)

Decays and CP violation

E-761	Vorobyov, Leningrad	Hyperon decay studies (6/16)
E-773	Gollin, Princeton	Kaon decays that violate CP symmetry (4/12)
E-774	Crisler, University of Illinois	Search for new low-mass particles (dark matter) (4/7)
E-800	Johns/Rameika	Magnetism of the omega particle (4/16)

Heavy quarks

E-687	Butler, Fermilab	Production of charm and beauty particles by photons (8/58)
E-690	Knapp, Columbia	Production of charm and beauty particles by hadrons (5/21)
E-760	Cester, Torino, Italy	Charmonium states (7/59)
E-771	Cox, University of Virginia	Beauty production by protons (9/68)
E-781	Russ, Carnegie-Mellon	Large-X baryon spectrometer (7/26)
E-789	Kaplan/Peng, North Carolina University	Production and decay of B-quark mesons and baryons (4/17)
E-791	Appel/Purohit, Princeton	Detailed study of the properties of beauty and charm particles (6/40)

Hard collisions and quark chromodynamics

E-672	Zieminski, University of Indiana	Study of very energetic "jets" and dimuons (7/28)
E-683	Corcoran, Rice University	Production of jets by high-energy photons (9/33)
E-704	Yokosawa, Argonne Lab	Experiments with a polarized beam of protons (16/50)
E-706	Slattery, University of Rochester	Direct photon production (9/75)

		COLLIDER PHYSICS

E-710	Orear/Rubinstein, Cornell	Total cross section of proton–antiproton collisions (6/18)
E-713	Price, Argonne Lab	Highly ionizing particles, e.g., monopoles (2/3)
E-735	Gutay, Purdue	Search for an unbound quark and gluon plasma (7/52)
E-740	Grannis, Stony Brook	Construction of a new detector for proton–antiproton collisions (20/124)
E-741	Shochet/Tollestrup, University of Chicago	Collider detector at Fermilab for proton–antiproton collisions (20/247)
E-775	Shochet/Tollestrup, University of Chicago	Upgrade of collider detector at Fermilab (20/247)

		OTHERS

E-466	Porile, University of Illinois	Nuclear fragments after high-energy collisions (3/7)
E-754	Sun, Buffalo University	Channeling tests (passage of particles through crystals) (4/8)

*Numbers in parentheses denote total number of institutions and physicists, respectively.

DARK MATTER, NEUTRINOS, AND CLOSURE OF THE UNIVERSE

We discussed in Chapter 6 the evidence that in addition to galaxies and clusters of galaxies, the universe has some nonshining, or dark, matter. We have mentioned that the orbits of stars around galaxies show that galaxies have masses about ten times what we can estimate from the matter we can see. Thus, galaxies have massive, nonluminous halos surrounding them. But even allowing for the dark halos, if $\Omega = 1$, we find that 90% of the mass of the universe is unseen and not made out of normal stuff. Could it be made of one of the supersymmetric particles now being searched for in colliders? Could it be planetary-mass black holes? Could there be some new particle as yet undreamed of by theorists? Accelerator experiments are searching for new stable particles that could be this dark matter.

One candidate for dark matter is the neutrino, *provided* at least one of the three known neutrinos has a mass of at least 20 eV. Now this is an exceedingly low value; the lightest known non-zero mass particle being the electron, with a 500,000-eV mass. Efforts to measure the mass of any of the neutrinos have intensified in the late 1980s. Direct measurements made on electron neutrinos emitted in radioactive decays of tritium have been inconclusive, although these experiments and the data we have

The neutrino detector at CERN. Such detectors typically contain a great mass of material (mostly glass) in order to increase the probability of neutrino collisions. The glass is in the form of many thin plates, which are intimately interspersed with particle detectors. This apparatus is capable of recording hundreds of thousands of neutrino collisions in a year with a great deal of detail. Similar detectors are operating at Fermilab and in the Soviet Union.

gathered from Supernova 1987A probably require the mass of the electron neutrino to be less than 20 eV. A subtle variation on an experiment to detect mass in the neutrino system depends on the possibility of one form of neutrino converting to another. It turns out that this can only happen if both particles have a non-zero mass. These "oscillations," of one species of neutrino into another, have been the subject of many experiments because the pay-off is large: Identification of mass in the neutrino could provide the mass needed to close the universe and account for the dark matter. Imagine the following situation: At 3 A.M. some midwinter morning, in a counting trailer at Fermilab, a tired graduate student on the graveyard shift reads the data stream from his neutrino oscillation experiment and discovers that more than enough mass is there. This one individual, out of the some 5 billion inhabitants of the planet, would know that the universe is ordained to cease its expansion, and eventually reverse into the big crunch! Heady stuff.

THE ELECTRON–POSITRON COLLIDERS

An immense variety of research is being undertaken by the machines listed in the table on page 193. One project has electron–positron machines exploring the 100-GeV domain, with the energy expected to be raised to 200 GeV at CERN by 1992. A major thrust here is to produce large numbers of Z^0 particles. Electron–positron collisions at a total energy exactly equal to the mass of the Z^0 will produce copious Z^0 particles. The CERN Large Electron–Positron (LEP) machine expects to produce more than 100,000 Z^0s per year. These all decay in 10^{-17} seconds, that is, before having traveled more than a fraction of a micron (10^{-6} meters). There are a huge number of decay modes, and all will be interesting. Given enough Z^0 particles, statistical analysis of the way in which they decay into neutrinos should establish whether there is a fourth kind of neutrino, and so a fourth generation of basic particles. According to big-bang nucleosynthesis theory, the measured abundance of helium makes a fourth neutrino very unlikely, and so the Z^0 research will be a crucial test of this important cosmological prediction. The LEP accelerator will be furnished with four collider detectors, each designed to amass definitive data on the decay properties of the Z^0. These detectors take between 5 and 8 years to build and represent the collaborations of 300 to 400 scientists. We have already commented on the most ambitious LEP detector, L3.

At Stanford University, another version of an electron–positron collider is scheduled to begin operation in 1989. Although this collider's energy is the same as at CERN, the accelerator technology is completely different, based on Burton Richter's idea that ultimately, head-on collisions of electrons and positrons will take place in linear, rather than circular, colliders. The SLAC Linear Collider (SLC) is, partially, a prototype of this idea.

EXTENDING THE STANDARD MODEL

Much of the research to be done using the accelerators listed in the table on page 193 is programmatic, filling in essential details, specifying the many parameters of the standard model with greater precision, sharpening our knowledge of the properties of quarks, leptons, gauge bosons, and so forth. However, in two kinds of experiments the main thrust is the search for extensions of the standard model. These extensions are attempting to find the true road to grand unification and to the ultimate theory of how the universe works, which we call the theory of everything.

The straightforward way to extend the standard model is to raise the accelerator energy enough to produce particles more massive than could have been produced by existing accelerators. This approach is exemplified by Fermilab's Tevatron.

Starting up in 1987, the Tevatron proton–antiproton collider at 1.8 TeV is the highest-energy accelerator in the world and will hold this record until the Superconducting Supercollider (SSC) or the Large Hadron Collider (LHC) comes into operation toward the end of the 1990s. Even when we take into account the constituents in the proton, the Tevatron will be exploring the unknown energy-mass domain of 200–500 GeV. Some of the candidates for discovery are the top quark (remember that we still need to finish the standard-model table), SUSY particles, fourth-generation quarks and leptons if they exist, or even traces of the mysterious Higgs particle.

CERN's proposed LEP II, the Soviet's UNK, and the ultimate machine for the 1990s, the Superconducting Supercollider, are also higher-energy machines. A European version of the SSC (LHC, the Large Hadron Collider) is also being considered. A more subtle search for the new physics of extremely high energies can be carried out by looking with great precision at the decays of lower mass particles. Just as proton decay, if found, would be evidence for GUT-energy physics, we would learn much about physics at extremely high energies if we found disintegrations of neutral kaons that violate the conservation of leptons, for example, into a positron and a muon. In specific GUT theories, the probability of such a decay can be estimated; it typically would be 10 billion times less probable than the common decay mode of kaons into pions. The experimental challenge is to amass many tens of billions of kaons, and design detectors to identify the positron–muon decay beyond any doubt whatever. Here again much work goes on.

THE MACHINE FOR THE 1990s: THE SUPERCONDUCTING SUPERCOLLIDER (SSC)

In 1983, the U.S. community of particle physicists proposed to the Department of Energy that it fund a new accelerator, the SSC, designed to address the most important issues in particle physics. Two spectacular aspects of the SSC were evident. First, its energy would be achieved by colliding protons against protons at a total energy of 40 TeV. Because protons are relatively easy to accelerate in large numbers, the collision rate would be very high: 50 million collisions per second would take place in each of up to six colliding regions. The second spectacular aspect was the cost: some $4 billion to be spent over eight years or so.

The sweep of the proposal, its cost, and the fact that a nationwide search was undertaken for a suitable site created a media event that occupied national front pages after the president of the United States gave his support to the project in January 1987.

The process of convincing the president, a sufficient segment of the scientific community and, indeed, the general public took from 1983 to 1987, a period noted for increasing concern with national deficits. (This is a separate story, no doubt suitable for a Ph.D. thesis by a new historian of science.) Then two days after the 1988 presidential election, the town of Waxahachie, Texas, was chosen by the Department of Energy as their preferred site.

The technology of the SSC was a straightforward extension of Fermilab's superconducting Tevatron, but with improvements in the niobium–titanium superconducting wire leading to a better magnet design with a magnetic field 50% higher than the Tevatron. With this new magnetic field and the desired energy of 20 TeV, the SSC would have needed a radius of 10 kilometers, if it were simply round. Because of the long straight sections required for collision regions, the actual circumference will be 83 kilometers.

(a) A possible layout for the Superconducting
Supercollider, which will be about
85 kilometers in circumference. In (b), the
beam pipe and its magnets are shown
oversize for clarity.

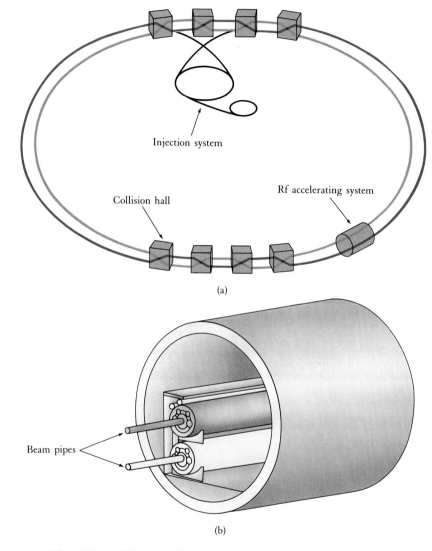

Injection system

Collision hall

Rf accelerating system

(a)

Beam pipes

(b)

The SSC will be installed in a tunnel deep enough underground
not to disturb normal surface activities. In the tunnel, two superconducting rings of magnets will carry protons around in opposite directions. In
the long straight sections, some six collision regions will be located.

It is expected that a large variety of particle detectors will be constructed in the SSC to observe the head-on collisions at a total energy of
40 TeV.

To understand the scientific drive for the SSC we must recall the
discussions in Chapter 6 on the grand unification quest and especially on
the Higgs field problem. We have seen that in particle physics, the Higgs
field and associated Higgs particles first appeared as a concept to unify

Magnet dipole for the Superconducting Supercollider. The superconducting cable winding is shown being assembled at Brookhaven National Laboratory. The coils will be placed around the 7.5-centimeter-diameter vacuum pipe.

the weak and electromagnetic forces. Further theoretical work indicates that some kind of symmetry-breaking field is needed at each stage of unification: strong with electroweak, gravity with the GUT theory. Also, the Higgs phenomenon is an essential ingredient in the inflationary version of big-bang cosmology.

Now, the standard model (SM) cannot deal precisely with the Higgs idea, although it does predict an approximate upper limit to the mass of the Higgs particle. The SM agrees with an enormous body of experience in a consistent manner, but problems arise when we try to extend the model to energies far above our experience. Here we find that the mathematics goes sour; quantities that should be measurable come out infinite. How can we save the valid part of the theory that applies to experiments below 100 GeV without predicting nonsense at high energies? The answer theorists have proposed is that there must be a Higgs-type particle, with a mass not much more than 1 TeV.

Theorists often try to hedge their speculations, but here no hedge was possible. The prediction was not precise: The mass of the particle could be 1.5 TeV perhaps, but already the extended SM math was becoming shaky. Suppose there is no such Higgs particle? Theorists then noted that something else, some brand-new, as yet undreamed-of effect must intervene to preserve the consistency of the standard model. And this new effect must be observable in the energy domain up to 1 TeV. So, what sold the idea of SSC was the Columbus metaphor: If Columbus didn't find India, he'd find something else of equal or greater interest.

Experimental physicists began paper design of detectors to work at SSC energies and collision rates and began to write complete programs to simulate the production of Higgs particles in these imaginary detectors. In principle, finding the Higgs particle is straightforward. Using the SM, theorists estimate that at a high enough energy, two protons can produce the Higgs boson H^0 (plus lots of junk). The Higgs boson is unstable and would decay almost immediately into a variety of objects:

$$H^0 \rightarrow Z^0 + Z^0$$
$$\rightarrow W^+ + W^-$$
$$\rightarrow t + \bar{t}$$
$$\text{etc.}$$

It turns out that this particle prefers the most massive decay products consistent with conservation of energy. (Here we assume that the mass energy of the Higgs particle is greater than twice the mass of the Z^0.) The detector does not see the Z^0, because it too lives a short time and has many decay modes, such as:

$$H^0 \rightarrow Z^0 + Z^0$$
$$ \longrightarrow e^+ + e^-$$
$$ \longrightarrow e^+ + e^-$$

The total reaction then would have a generous spray of pions and kaons, and four electrons, which can be flagged by detector elements sensitive to them.

Measuring the momentum of each electron gives the data needed to calculate the mass of the parent. If the electrons were derived from a variety of background sources of electrons, the result from, say, 100 events would be an almost random set of numbers. If, among this background, there were, say, twenty true Higgs bosons, there would be twenty numbers clustered closely around the mass of the particle. Such a

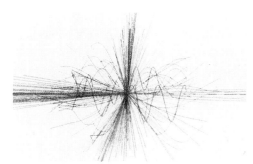

A computer simulation of a proton–proton collision in the Superconducting Supercollider. In this example, a Higgs boson has been produced, together with many other particles. Simulations like this enable physicists to design detectors that will extract the most relevant information from the data.

mass "peak" is revealed by plotting the number of events against the mass calculated for the parent, as we have already seen Ting do to find the J/ψ particle on page 71. This peak can easily show up even with the background of five times as many sets of four electrons.

The difficulty is that a 1-TeV Higgs particle would be produced in only one out of every 10^{10} collisions. Further, the fraction of such particles decaying to two Z^0s may be less than 100%. If, for example, it is only 30%, two out of three Higgs particles are not available for this test. Also, the probability that both Z^0s decay into electron–positron pairs is only a few percent. Finally, no apparatus is entirely efficient; if one of the electrons is lost in a crack or wrongly identified, we lose the event. Taking all this into account, we need to produce about 10,000 Higgs bosons to be sure of a discovery. This is why a collision rate of 50 million per second is important. It will take more than a year of running to collect enough data to get an observable Higgs peak if the mass is close to 1 TeV. This would be great, but success is never assured because of the interference of huge numbers of irrelevant collisions; everything depends heavily on the art of detector design.

In the summers of 1984, 1986, and 1988, intensive work was carried out by hundreds of physicists on the detector problems associated with SSC. Let's compare the SSC with the Fermilab Tevatron, working near 2 TeV. Each interesting SSC collision will produce three or four times as many particles as at the Tevatron. In contrast to the leisurely collision rate at the Tevatron, which allows several microseconds for the detector to "digest" an event, the SSC detector gets only 10 nanoseconds (10^{-9} seconds)! Finally, the average energy of the emerging particles is roughly ten times higher. Small wonder that the SSC global detector will weigh 50,000 tons and have over 1 million channels of data. The technology of this almost ultimate microscope will push the state of the art in time, space, and energy measurements to a new level of sophistication.

There are many other approaches to the Higgs problem, and one of the SSC's main goals will be to illuminate this most puzzling aspect of the inner-space/outer-space convergence.

SUPERSYMMETRY

In Chapter 6 we discussed supersymmetry, which predicts a rich spectrum of particles, some of which *must* have masses below 1 TeV *if* the SUSY theory is correct. Again, the SUSY theory, extending the standard model, is specific enough that it predicts the probability of proton–proton collisions producing, for example, a gluino (SUSY partner of the gluon) or a squark (SUSY partner of a quark). The unknown quantities in these theories are the masses, and the theorists will cheerfully estimate how many gluinos will be produced per collision at a given mass. At SSC energies (or even at Tevatron energies), we can then estimate the discovery potential for masses from 50 GeV to 1 TeV. To actually make the discovery, we must look at the decay modes, for example, a gluino decays to quark plus squark plus photino. What does the detector see? Quarks will emerge with considerable energy; with characteristic shyness, the quark will convert into a narrow jet of hadrons—pions, kaons, and so forth. The squark will also appear in the detector as a jet of normal particles. The photino has only a small probability of being detected at all, and will escape the detector, giving rise to missing momentum. That is, when all the particles are measured and the total momentum added up, the difference between that total and the zero value required for conservation of momentum will indicate the direction and momentum of the escaped photino.

Thus (if the detector is good enough), the photino momentum (and energy) can be deduced and added to the two jets (momentum and energy) to calculate the parent rest mass. If a specific mass shows up enough times, we again obtain a *mass peak*, and if the peak can be distinguished from the background, we have established the existence of a new object at the mass indicated by the peak. More detailed studies of the jets will eventually confirm the hypothesis that we have witnessed the birth of a gluino.

COLLIDER DETECTORS

The quest for new discoveries by machines in the 1990s clearly depends on the quality of their collider detectors. Physicists have now had much experience in building these, going back to the early SPEAR electron–positron machine at Stanford, and proceeding to the CERN UA1 and UA2 detectors that discovered the W and Z gauge bosons as well as Fermilab's CDF and Dzero detectors at the Tevatron. These are immense devices, costing over $50 million, taking five to eight years to build, and requiring the efforts of hundreds of physicists and engineers.

The detectors surround the interaction point where head-on collisions take place. In a world of pure mathematics, they would be spherical, with holes only for the vacuum pipes carrying the beams in and out. In practice, their forms depend on the cylindrical shape of the machine and on the density of radiation expected to emerge from the collision.

This portion of the Collider Detector at Fermilab detects particles emitted from the proton–antiproton collisions toward the forward (proton) direction. The central detector has been rolled out, and the beam pipe appears between the calorimeters (red rectangles). Behind these are circular magnetized iron toroids, which detect high-energy muons and measure their energy.

The goal is to measure all you can about each event. An observer would like to detect particles emerging at any possible angle, with enough resolution to be able to distinguish between closely spaced tracks. The observer would also like to identify each particle if possible, measure its momentum by its curvature in a magnetic field, and measure its energy in calorimetric devices. As we noted above, missing energy and momentum are also important; so there should ideally be no cracks by which particles can escape detection other than by their unique ability to pass through detectors without leaving signals, as neutrinos or photinos can.

Detectors can be designed to identify electrons and photons from the characteristic electromagnetic "showering" they cause in thin lead plates. The shower says we've had a particle that interacts purely electromagnetically. The tracking tells whether the culprit was charged or neutral. Even more discriminating means are available to pick electrons out of a spray of other particles. Muons can be detected, because they pene-

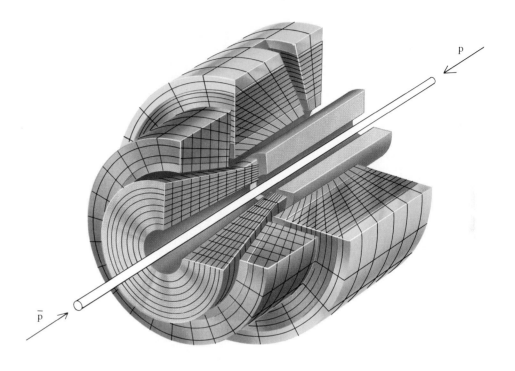

A schematic of a collider detector, showing the segmentation of the tower structure. The tracking chamber (green region) registers points on the trajectory of the particle with great precision. Leaving the tracking chamber, the particle encounters a series of thin lead plates separated by scintillation counters (blue region). Electrons are totally absorbed here, and their energy recorded. Surviving this, the nonelectrons encounter a series of iron plates spaced by devices that measure the total energy carried by the particle (red region), whether charged or neutral. Finally, any particle able to penetrate to the detectors in the orange region will be identified as a muon.

trate many meters of steel without interacting. Hadrons have strong interactions in the calorimeter sections, giving rise to nuclear cascades.

Combining these characteristics, we arrive at the idea of a *layered* detector. In such a detector, as we proceed outward from the collision point, we have a few centimeters of vacuum, a very thin vacuum-chamber wall, and then, usually, two layers of tracking. The first layer is high-precision tracking, which can deduce whether the track originated in the collision or from a secondary process, to an accuracy of 50 micrometers (10^{-6} meters) or better. This would detect very short-lived particles. The second layer is usually one or two meters thick, and uses a MWPC system to register the trajectory. Usually there is a magnetic field here. The next layers use electromagnetic detectors, followed by hadronic detectors. Survivors are either muons, which are tracked at the outside, or noninteracting particles, e.g., neutrinos.

In addition to layering, there is also towering; that is, so many tracks emerge from a high-energy collision that we need to separate the detector into many small angular towers, designed so that the probability

The three-story-high Collider Detector at Fermilab. The black arches removed to the sides contain modules of the central calorimeter. Photomultipliers and electronics inside the calorimeter measure particle energies from collisions. The detector provides 70,000 channels of data to the dedicated CDF computers.

for two tracks in a given tower is small. The figure on the facing page illustrates the layered towers in the Collider Detector at Fermilab.

It is important to realize that a typical collider detector may have 100,000 wires, each one carrying information from the hardware end to an array of electronics that is called the Data Acquisition System. All of these channels of data must be organized, checked, and eventually processed in order to present the experimenters (or, more likely, their computer) with a decision as to whether to record the event on magnetic tape or discard it. Since only one event in many tens of thousands (or even millions) is interesting and the maximum speed of recording is relatively slow (a few per second) these electronic decisions are an important part of the art and science of detectors.

TELESCOPES ON THE SURFACE OF THE EARTH

Telescopes in their most general definition are instruments designed to detect and measure signals from outside our planet. As such, they fall into three general categories: those used from the surface of the Earth, those used from space, and those used underground. For the 1990s, each of these dwelling places for telescopes will be used more fully and productively than had ever been dreamed possible. We will now examine each of these types of telescope placement, discussing the kinds of instruments that are expected to be developed for use in the 1990s and what we hope these instruments will find. Of course, modern telescopes are a far cry from the tube Galileo squinted through to see the moons of Jupiter. In the 1990s almost the entire electromagnetic spectrum is surveyed by specifically tailored instruments. In addition, charged particles, neutrinos, and gravity waves arriving from outer space are all included in the objectives of these new telescopes.

Traditionally, telescopes have been fastened to the surface of the Earth to look out through our atmosphere at the "light" coming from outer space. Initially, the light was standard optical light. But as we discussed in Chapter 5, this searching for extraterrestrial light eventually extended into the radio wavelengths. The Earth's atmosphere, gloriously transparent (usually) to visible light, does terrible things to other wavelengths of the electromagnetic spectrum; so the surface of the Earth is not a very good place to build telescopes to look into large parts of the infrared spectrum, the ultraviolet, the X-ray, or the gamma-ray regions. Actually, life on Earth is lucky to have this shielding atmosphere, because people and other life forms do not respond well to intense doses of X rays and gamma rays.

In the 1990s, two major developments for the inner-space/outer-space connection will take place utilizing ground-based telescopes. One will be the operation of extremely large telescopes and the other will be the development of telescopes dedicated to cosmology (i.e., not used for any other purpose).

One major development will be the deployment of extraordinarily large (at least 8 meters across) ground-based telescopes, to catch the very weak optical light from distant objects. By going to very large optical telescopes, we can gather light from extremely faint objects in space in a relatively short time. When these giant mirrors are pointed at very distant objects, they can pick up and add together each of the individual photons striking the mirror. The diameter of the mirror contributes in two ways to the information collected. One is in the amount of light gathered, which is proportional to the area of the mirror. The other has to do with the wave nature of the light, which tends to blur the image of a point of light, for example, a star. This blurring is known as the *diffraction limit* of the optical system. The laws of physical optics teach us that the radius of the blur is proportional to λ/D, where λ is the wavelength of light and D is the diameter of the mirror. Thus, the larger the diameter, the smaller the blur. (Incidentally, this also holds for lenses, even your own camera lens.) Among other causes of blurring are imperfection in the shape of the mirror, any shimmering due to thermal gradients in the atmosphere, and tiny vibrations of the mirror, such as those due to passing subway trains, airplanes, or microseismic disturbances. Collecting enough light to make out the existence of a star is essential, but minimizing the blur is also important, since blurring can obscure the fact that, say, two stars are close together or that a source is actually a distant cluster of stars. The sharpness of the image is called *resolution*. For telescopes operating at optical wavelengths, "seeing" through the atmosphere is more important than controlling the diffraction limit, but for infrared, microwave, and radio waves, the diffraction limit is critical. Furthermore, seeing is affected by such things as the building housing the telescope and the geography around it. Astronomers hope to operate the giant new telescopes in both optical and infrared wavelengths.

The giant telescope mirrors can be fabricated in a variety of ways. One is to cast a single large disk. However, such disks, if scaled up from the technology of Mt. Palomar or Kitt Peak, would be exceedingly massive hunks of glass, slowly sagging under their own weight as time passes, thus making their images unstable. And, because their mass is so high, the structures to guide them would be very expensive. The Soviet Union attempted to make a 6-meter telescope from a huge piece of glass. This telescope continues to be plagued with problems, and has not been nearly as useful a scientific instrument as the smaller 5-meter Palomar Observatory or the even smaller 4-meter Kitt Peak Observatory. Furthermore, large masses of glass retain differences in heat for a long time, creating convective air patterns near the surface of the mirror. These air patterns make for poor observation.

What has enabled people to contemplate 8-meter single mirrors is a new technology of casting mirrors using a glass honeycomb. The mirrors are quite thin and lightweight, and require a strong but lightweight structure to support and guide them. The mirror itself is almost flexible since it is so thin, and its support, rather than the glass itself, enables it to maintain its shape. Since it contains relatively little glass, the mirror can rapidly come to thermal equilibrium with its surroundings. There

The 3.5-meter "honeycomb" mirror casting for the ARC telescope to be placed at Sac Peak in New Mexico. This is the first large telescope to use the new lightweight honeycomb mirror technology. ARC is the acronym for the Astronomy Research Consortium, which consists of Princeton, the University of Chicago, the University of Washington, Washington State, and New Mexico State.

are thus fewer convective air patterns, thereby improving observation. The large honeycomb mirrors achieve their shape by having the honeycomb jig (the mold used to cast the mirror) and its molten glass rotate so that the glass flows to the outside. This results in an almost perfect parabolic shape that requires minimal polishing. The old solid hunks of glass required polishing to remove the entire parabolic interior.

In addition to having single, large mirrors, designers are also discussing making 10-meter or larger mirrors out of combinations of smaller 4-meter mirrors held closely together. The mirrors would be kept in focus with one another by use of lasers bouncing off the surface and giving feedback to a computer on the relative position of the mirror. The smaller mirror supports could be moved rapidly enough to keep the mirrors in focus. The California Institute of Technology/University of California Keck Telescope now being built on Mauna Kea in Hawaii uses this technique with smaller hexagonal mirrors. A variant on this technique has already been used in the multimirror telescope operated on Arizona's Mt. Hopkins by the University of Arizona and the Harvard-Smithsonian Center for Astrophysics in Cambridge, Massachusetts. However, in the multimirror telescope, each mirror is itself a full parabolic mirror, whereas in the segmented Keck Telescope, each mirror is a segment of the larger 10-meter parabola. This latter plan may make for better images, but has the difficulty of requiring the polishing of separate segments, none of which is a full parabola.

Another option is to put two or more 8-meter mirrors beside each other and do interferometry between them. *Interferometry* adds together

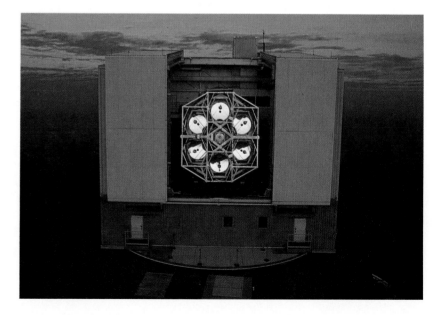

The multiple mirror telescope on Mt. Hopkins in Arizona, the first major telescope to use more than one mirror to obtain an optical image.

The planned Columbus telescope will use interferometry to combine light rays reflected off two 8-meter mirrors.

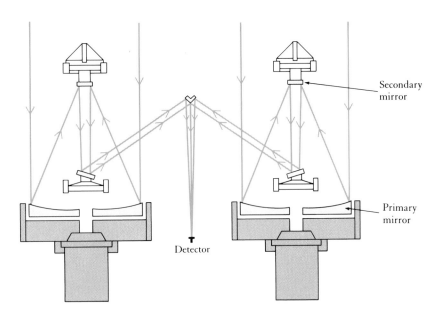

the light waves coming to each of the separate telescopes. Such addition must be done with great care to keep track of the phases of the incoming light waves. (The word *interferometry* comes from the concept of "interference" between separate waves.) Such an array would have the resolution of a telescope equal in size to the distance between the farthest parts of the two telescopes; that is, two small mirrors 100 feet apart would have the same resolution as a 100-foot mirror (but not its light-gathering ability).

This kind of interferometry has worked very well with radio telescopes operating across the entire United States. In effect, the distance D in the resolution expression is extended to 3,000 miles! However, it is much more difficult to get optical light waves than it is to get radio waves into phase with each other so that they can be added properly. New techniques have proven that this light addition can occur in the infrared, and astronomers are pushing the techniques on into the optical regions. Two 8-meter telescopes separated by a distance of 5 meters would have a resolution equivalent to that of a 21-meter telescope. The light-gathering capability of such an instrument is that of the sum of the areas of the two 8-meter disks, which would be equivalent to a telescope 11 meters in diameter. Such a project is being developed by a collaboration between the University of Arizona, Ohio State University, and the University of Florence and possibly the University of Chicago in a venture known as the Columbus Project. The European Southern Observatory, which is funded by most of the nations of Western Europe, much as

A model of the four 8-meter telescope installation being built by the European Southern Observatory in Chile.

CERN is, is building an array of four 8-meter telescopes in Chile, so that the southern sky can also be observed with extraordinarily big "light buckets."

The advantage of these giant telescopes, as mentioned before, is that very distant and faint objects can be observed, and less faint objects can be observed in a shorter time. Therefore, measuring the atomic spectral lines necessary to determine redshifts can be carried out rapidly on the nearer galaxies.

The disadvantage of these large telescopes is twofold. First, they have to look through the atmosphere and are subjected to clouds, weather, the blocking of ultraviolet and infrared light, and the distortions of the air. Second, these telescopes are so expensive that they require the combined financial resources of large consortiums of astronomers in order to be built. The consortiums tend to be so large that the astronomers involved cannot agree on single observing projects, and so they plan on many projects. This is fine and exciting, but it means that telescope time must be shared, and that no single project will get more than a few nights per year. Unfortunately, some very important projects require far more than a few nights of telescope time per year. This leads us to the conclusion that we need another class of new telescope for the coming decade.

The second major development for ground-based optical astronomy will be dedicated 4-meter telescopes. These will devote all of their time to a single problem of cosmological significance. Four-meter-class telescopes are roughly equivalent to the very best telescopes we have operating in the 1980s. However, because these are the best telescopes in the 1980s, their time is divided up in the way that the time of the future 8-meter-class telescopes will be.

There are certain projects that require hundreds of nights of observing. An important example is the mapping of the sky in three dimensions. To do this, we need to know the velocities at which the galaxies are receding, so that we can use Hubble's law of expansion to find their distances as well as their positions on the celestial sphere. To find a galaxy's velocity requires measuring the apparent shift toward long wavelength (i. e., redshift) of the atomic lines in the light emitted by the receding galaxy. To get redshifts requires looking at the galaxy not only long enough to see that it is emitting light, but also long enough to be able to decompose the light into its component wavelengths and to identify atomic lines. Three-dimensional maps of the sky that contain galaxies at very large distances can be used to discover the large-scale structure of the universe, that is, how galaxies are clustered; and their distribution in sheets, voids, and foam; among other things. As we mentioned in Chapter 6, whatever the universe's structure, it was generated, at least in part, by phase transitions near its birth. Thus, by understanding the structure, we may be able to understand the fundamental forces that created the universe.

The preliminary maps that have been carried out have taken years to accumulate, using the few dedicated telescopes that do exist. But these telescopes have been of the 1- to 2-meter class, not the 4-meter class, and don't have enough light-gathering capability to get redshifts on very distant objects. As a result, surveys so far have been able to map galaxies out to only a few hundred megaparsecs. Some surveys have taken another approach, looking in one direction for a very long time, seeing all the galaxies in one direction out to about a thousand megaparsecs. This map then exposes one single, narrow, "pencil-beam" direction. So far, the results have been spectacularly surprising. Instead of a more or less homogeneous distribution of galaxies, occasionally enlivened by clusters, huge voids knitted together by threadlike arrays of galaxies have emerged. The patterns are as intriguing as they are beautiful, but only a tiny slice of the sky has been mapped in this way.

Dedicated 4-meter-class telescopes will be able to change this. They will generate surveys that go very deep, with redshifts for millions of galaxies, thus charting the structure of the universe. Eventually these telescopes will answer questions such as: What constitutes clusters? Where are voids? Are things on the large scale really distributed in sheets, filaments, sponges, and so forth?

RADIO ASTRONOMY

In addition to these two major developments in optical astronomy, radio astronomy will be advancing with new interferometric techniques at the *very long baseline array* (VLBA) that will allow extraordinarily high resolution of very distant objects. This array will consist of large, high-quality radio telescopes distributed over North America (see the figure on this page). It will have a power of resolution equivalent to a telescope that is 3,000 miles across, although its light-gathering power will be equal only to the sum of the areas of the individual antennae. With such high resolution, we will be able to look at the cores of galaxies near the time of their birth.

Another effort in ground-based radio astronomy is to look at details of the microwave background radiation. It is true that to see some

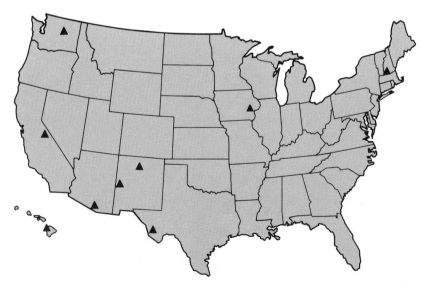

As the map shows, the unified array of radio telescopes comprising the VLBA (very long baseline array) covers most of North America. It will yield a resolution equivalent to a telescope with a 3,000-mile-diameter-mirror. However, the signal gathering power will be proportional to the sum of antenna areas.

parts of the microwave spectrum, astronomers need to get above the Earth's atmosphere. However, much of the microwave spectrum is still quite visible from the ground. Newer, more precise radio antennae are being designed and developed to look at this microwave background radiation. They will try to detect slight variations in the temperature from one direction to another. Such variations are called *anisotropies*. As we discussed in Chapters 5 and 6, the microwave background gives us a powerful probe of the era just before galaxy formation. The bumps and wiggles in the structure of the universe at that epoch eventually grew into galaxies and clusters of galaxies. Unfortunately, today there has been no detected variation in the microwave background on scales that might be appropriate to galaxy formation. It is very important to make these experimental measurements more precise, to see at what point some small lumpiness begins to appear in the otherwise smooth background. Already the limits on anisotropies show no fluctuations down to a few parts in 100,000. To reduce the background for these microwave antennae, people are proposing an installation at the South Pole, where the air temperature is low and a single "night" can last for months.

COSMIC RAYS AGAIN

Until the 1960s, cosmic rays were effectively used to explore the properties of the subnuclear world in competition with particle accelerators, enabling the discovery of many particles, including the positron, the muon, the pion, and the kaon (charged *and* neutral), as well as the lambda hyperon. However, study of the detailed properties of each of these particles awaited accelerators that could produce the particles copiously and in a controlled manner rather than waiting for random interactions coming from space. This was an interesting aspect of the inner-space/outer-space story. Gradually, as accelerators became more powerful, the study of cosmic rays began to emphasize an understanding of the astrophysical sources of the radiation arriving on our planet. By the 1980s the subject had a new name, high-energy astrophysics, and it required major detector facilities.

Not all scientists have given up on the particle-physics aspects of cosmic radiation, and as we talk about exceedingly high energy accelerators, some scientists again turn to the cosmic rays to see if they can get some inkling of what might be found at these very high energy accelerators. Because the energies that we are now talking about with accelerators are so high, the intensity of cosmic rays at those energies is correspondingly low, and the cosmic-ray detector experiments must be-

Part of the Fly's Eye cosmic-ray detector in Utah. This multiple mirror array detects very high energy cosmic rays when they interact with the Earth's atmosphere. The Fly's Eye can see such interactions over a 1,000-square-kilometer section of the sky.

come enormous to have a collecting area big enough to see anything. For example, in the deserts of northwestern Utah is a cosmic-ray detector known as "The Fly's Eye," which scans 1,000 square kilometers of the sky to look for cosmic rays hitting the Earth's atmosphere.

The Fly's Eye detector consists of two very large arrays of parabolic mirrors, all pointing at the sky. Each of these mirrors has a detector attached, so that a flash of light seen by that mirror can be recorded. In this way, the mirrors are mimicking the multitude of lenses of which a fly's eye is made. The purpose of the Fly's Eye is to look for flashes of light in the atmosphere of the Utah night sky. These short-duration flashes of light are caused by cosmic rays hitting the Earth's atmosphere. The higher the energy of the cosmic ray, the bigger the flash of light, and the longer its light trail. These detectors are calibrated so that the energy of the incident cosmic ray can be calculated.

This experiment has been operating for several years, and has approximately 100 counts above 10^{10} GeV. Even these small numbers can be important and intriguing. For instance, the present data are being used to test theories about how these very energetic particles react with the photons of the 3-degree-Kelvin background radiation. The Fly's Eye is already operating, but major expansions of it are planned for the next decade. These should enable it to find places where particles are accelerated to extraordinarily high energies in space. Perhaps, in seeing how nature makes gigantic accelerators, we might be able to take advantage of these ideas in future terrestrial experiments.

At present, not only is the Eye being expanded, but new detectors are being added to look at the same patch of sky in other ways. For example, in addition to the detectors looking for optical light, there is also an array of detectors that will look for gamma radiation. This array will be able to decide if the particle hitting the upper atmosphere is a regular proton-like cosmic ray, or is perhaps an ultra-high-energy gamma ray that hits an atmospheric nucleus and makes a shower of particles in the collision. These gamma rays will be coming in a straight line from their sources, unlike the charged cosmic rays, which spiral around on the magnetic fields of the galaxy and thus lose any information about their direction of origin. If enough gamma rays are detected from a single location in the sky, we can work back and see what object emitted them. Already there are hints that certain objects in our galaxy can emit extraordinarily high energy particles, with energies greater than 1,000 TeV, far beyond the reach of even the Superconducting Supercollider. In fact, some cosmic rays have energies as high as 50 million TeV.

To better understand all the cosmic rays that will be hitting the sky over the Utah desert, another type of detector is also being added to the Fly's Eye. Planted underneath the Fly's Eye detector is a new array of detectors, shielded by a thick layer of earth from all particles except muons and neutrinos, which will still make it through and trigger the detectors. The association of muons and neutrinos will add much to understanding just where this energy is coming from. This experiment is a nice example of how techniques from particle physics, such as muon and neutrino detectors and gamma-ray detectors, are combined and applied to studying astrophysical objects.

GRAVITY WAVES

Another form of ground-based telescope is one that looks for gravity waves. Gravity waves are small oscillations in space-time itself. They may be coming to us from distant galaxies, as black holes gobble up material at the core of these distant objects, or perhaps as two neutron stars collide into each other after their death spiral. Gravity-wave detectors also look different from the other telescopes. They are likely to be either giant supercooled bars of metal or crystal, or a fancy laser interferometer that spans miles of high-tech vacuum and electronics, laid out in remote, level areas away from other disturbances. (This interferometer compares the laser light waves bouncing off masses that are miles apart.) As the space-time fluctuation of a gravity wave hits the detector, the minuscule oscillations of the bar or reflected laser light produce detecta-

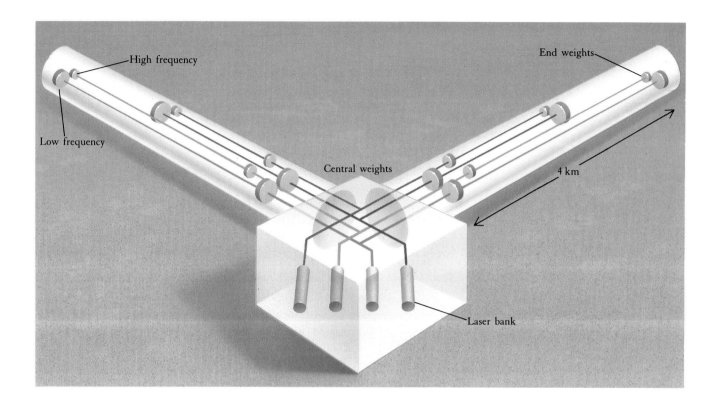

The proposed M.I.T.–Cal. Tech. gravity-wave detector would detect a tiny change in the distance between weights suspended in two steel tubes, each 4 kilometers (2.6 miles) long. The changes caused by gravity waves would be measured by comparing laser beams bounced off the weights in each tube. Gravity waves would induce movements which are detectable even when one one-thousandth the size of an atom.

ble signals. Two large interferometers are going to be built by scientists from M.I.T. and Cal. Tech., in California and at a yet to be named east coast location. At this time they are building and testing smaller laser interferometers in laboratories at the two schools.

SPACE INSTRUMENTS

To look at types of light that cannot penetrate the Earth's atmosphere, we must go above it. This has been done by putting telescopes on satellites that orbit the Earth. Until now, no large optical telescope has flown on such a satellite, nor has the cosmic background radiation been looked at from such a vantage point. Two of the most exciting telescopes of the 1990s will do exactly those two things: the Hubble Space Telescope (HST) and the Cosmic Background Explorer (COBE).

The first of these, the Hubble Space Telescope, is a 2.6-meter optical telescope that will orbit the Earth after being launched by the space shuttle. Once above the Earth's atmosphere and away from its distorting

The Hubble Space Telescope sitting in storage awaiting launch. Unaffected by the Earth's atmosphere, the 2.6-meter optical telescope will be able to see farther than a 5-meter telescope on the ground.

influence, the HST will have a very good view of objects in optical and even ultraviolet light. Thus, it will be limited only by the diffraction limit and the laws of optics, not by local atmospheric conditions. In order to be launched from the shuttle, it must be only 2.6 meters across. Thus, it will not even have the light-gathering capabilities of a ground-based 4-meter telescope, much less that of a ground-based 8- or 10-meter telescope. However, being above the atmosphere, it will be able to see somewhat farther than a comparable 2.6-meter telescope on the ground. In fact, it will have the same range of sight as a 5-meter telescope on the ground.

The main thing that the Hubble Space Telescope will do for cosmology is to establish the distance scale more precisely, because it will be able to see certain objects much farther away than they currently can be seen. As we mentioned earlier, the Hubble constant, which is a measure of the size of the universe, is uncertain by about a factor of 2. With the Hubble Space Telescope, we will be able to see the variable stars called Cepheids, not just those in the very nearby galaxies, but those in galaxies as far away as 20 megaparsecs in the Virgo Cluster. This will enable us to find a much more precise value for the distance to the Virgo Cluster,

....

COSMIC DISTANCE LADDER

Distance in parsecs	
Earth–Sun distance by radar ranging	$\sim 5 \times 10^{-6}$
Distance to nearest stars by parallax using Earth–Sun baseline	~ 1 to ~ 100
Distance to star clusters using properties of stars as ascertained from nearby stars	~ 100 to $10,000$
Distance to nearby galaxies using variable stars (Cepheids, etc.) whose properties were discovered from variables in star clusters	Few times 10^6
Distance to nearby clusters of galaxies (such as the Virgo Cluster) using a variety of techniques based on observations of objects in nearby galaxies—possibilities include: the brightest star in a galaxy; novas; supernovas; hot regions around bright stars; globular clusters	Few times 10^7
Distance to distant galaxies using properties of galaxies themselves such as: rotation velocities; brightest galaxies in clusters; supernovas	$\sim 10^9$

1 parsec is about 3 light-years or 3×10^{16} meters.

which is one of the cornerstones in the cosmic distance ladder (see table on the facing page). This should reduce the uncertainties in the Hubble constant from a factor of 2 down to accuracies of 20 or 30%.

The Cosmic Background Explorer known as COBE is the satellite that will be launched to look in detail at the cosmic background radiation. In particular, it will be able to measure the temperature of that background radiation much more precisely than has been possible from ground-based observations or even from high-flying balloons or sounding-rocket observations. COBE will be able to stay in orbit for many months and to take precise temperature measurements at various angular scales. It will also be able to see if there are any deviations in the temperature from wavelength to wavelength. Such variations in temperature would be indicative of things that might have happened in the background radiation near the time of galaxy formation. Variations in the temperature from one direction to another would be indicative of the origin of the fluctuations that eventually grew into galaxies and formed the structure of the universe. This is clearly one of the most important experiments that can be done in space.

A NASA artist's conception of the Advanced X-ray Astronomy Facility, scheduled to be launched in the late 1990s.

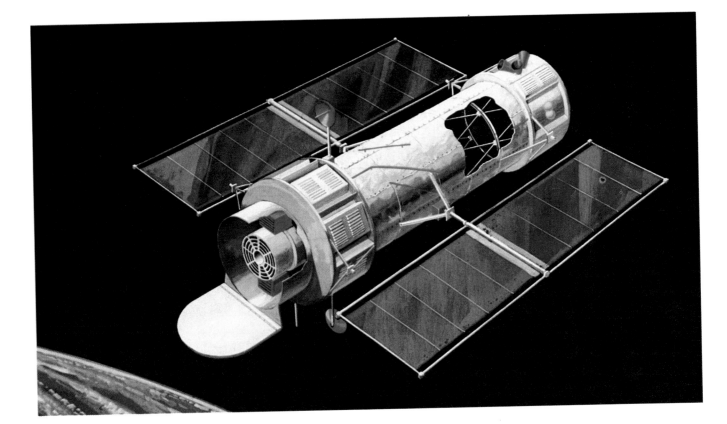

Another planned project of particular importance to the inner-space/outer-space connection is the Advanced X-ray Astronomy Facility known as AXAF. This won't be launched until the end of the 1990s, but it will do for X-ray astronomy what the Space Telescope will do for optical astronomy. It will be a semipermanent X-ray observatory placed in orbit by the shuttle. As with the Space Telescope, the shuttle will be able to go back and visit AXAF to put on new instruments, or make repairs. However, unlike the Hubble Space Telescope, there are no ground-based competitors for AXAF; X-ray astronomy can be done only from space.

From a cosmological viewpoint, one of the most exciting things that X-ray instruments can do is look at the emissions from galaxies at the moment of their birth. Most models for galaxy formation imply that some very exciting activity should have occurred right at their birth. A lot of this activity can yield X rays. A high-precision X-ray telescope in orbit, able to make repeated observations, can decipher the X-ray emissions at the birth of galaxies and help discover the mechanisms that cause galaxies to form and that determine their structure. Furthermore, many of the largest clusters of galaxies seem to have large quantities of X-ray emissions from the hot gas trapped by the gravitational pull of these clusters. Thus, by looking at how this hot gas evolves in ever-more-distant clusters, we might learn not only about galaxy formation but also about cluster formation.

Before the United States launches AXAF, the European Space Agency will be operating another X-ray satellite, ROSAT. This will have major instruments on it built by the Max Planck Institute in West Germany. ROSAT is not as large as AXAF, but it will be the most powerful X-ray satellite until AXAF goes up. ROSAT will be launched in 1990. With ROSAT, we will also be able to get some information about galaxy and cluster formation. Furthermore, ROSAT will be up soon enough that it should be able to see whether Supernova 1987A in the Large Magellanic Cloud left behind an X-ray-emitting neutron star. Neutron stars, when they form, are extraordinarily hot and emit most of their energy in X rays. When Supernova 1987A blew up, the only satellites looking at the X rays were low-resolution satellites put up by the Soviets (with West German instruments) and the Japanese. They gave us some exciting information, but it is unfortunate that the kind of high-precision work that was available in the 1970s with the Einstein Observatory was not in orbit at that crucial time. NASA did try to remedy the situation with rocket-launched instruments and balloons, but these did not go into orbit and so had only a brief time to make observations.

Another satellite that should be launched in the 1990s will be the Gamma-Ray Observatory, GRO. This will begin to do for gamma-ray astronomy what AXAF will do for X-ray astronomy and what the Hubble Space Telescope will do for optical astronomy. The Gamma-Ray

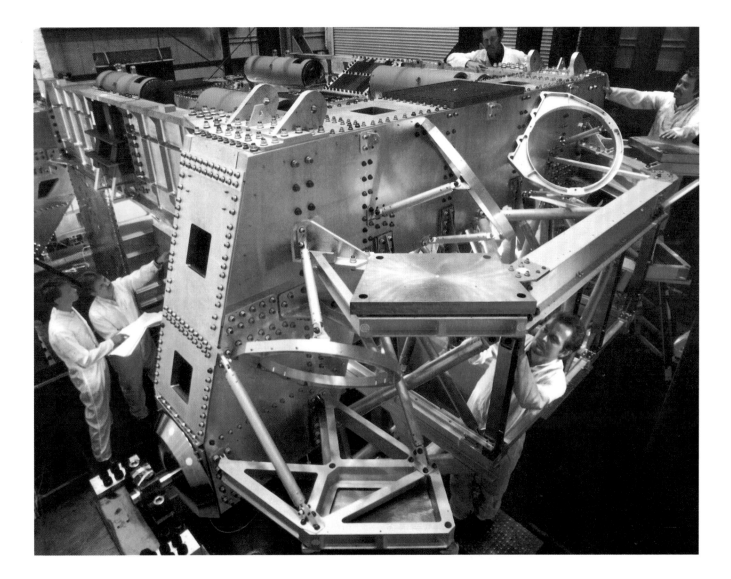

The Gamma-Ray Observatory under construction.

Observatory will be designed to look for gamma rays coming from objects in space. Gamma rays can be produced not only by very high energy accelerators in space, such as those that will be detected with the Fly's Eye observatory, but also by decays of radioactive elements freshly made in a supernova.

In fact, 100 days after Supernova 1987A exploded, light emission from it was dominated by energy coming from the decay of nickel 56 to cobalt 56 to iron 56. This radioactive decay chain gave off gamma rays. As long as the envelope of the supernova was still thick enough, those gamma rays were scattered by the material and reemitted in optical and infrared light and so did not reach Earth. However, as the envelope thinned out, more and more of the gamma rays escaped directly without scattering. These gamma rays were detected by a satellite—the Solar Maximum Mission—that was put up for a completely different reason. The Gamma-Ray Observatory will be a much larger gamma-ray satellite, and will be able to see many different gamma-ray lines from a variety of objects in space, as well as detect directions of many gamma-ray emitters. It will, for the first time, be able to map space in the gamma-ray wavelengths.

For cosmic rays, an exciting space project is known as ASTROMAG. It will be a large superconducting magnet carried on the space station. The magnet will bend the paths of the incoming charged cosmic rays and enable their composition to be precisely measured.

UNDERGROUND OBSERVATORIES

Perhaps the best physical examples of the new symbiotic relationship between particle physics and astrophysics are the underground observatories, in mines and tunnels—some with more than a mile of earth and rock above them, to shield them from everything that traditional telescopes look at. As we have already seen, these new "telescopes," deep underground, detect neutrinos. To do so, they must be both massive and thoroughly instrumented, so that the relatively small energies released by neutrino collisions can be measured. The most impressive of the underground observatories were actually built to look for proton decay, and it was these that detected the burst of neutrinos from Supernova 1987A, in the Large Magellanic Cloud, 170,000 light-years away.

The equipment in the Kamioka Mine in Japan can also detect neutrinos from the Sun, and it was used to check the solar neutrino results from a previous experiment operating in the Homestake Gold Mine in South Dakota. The Homestake experiment tank contained 100,000 gallons of cleaning fluid instead of water. The chlorine in the cleaning fluid absorbs neutrinos and changes chlorine atoms into radioactive argon, which can be detected. The Kamioka experiment confirmed the results of the Homestake experiment—that the Sun seems to be emitting neutrinos at a rate less than half of what theoretical astrophysicists had expected.

New underground experiments are being designed that will also be sensitive to neutrinos from supernovas and the Sun. Experiments being built in the Grand Sasso Laboratory, a large tunnel in Italy, contain 30 tons of gallium, which is also very sensitive to the neutrinos from the

Sun. An experiment in the Baksan Tunnel in the Soviet Union contains 50 tons of gallium. An experiment being developed in an iron mine in Canada will be built with 2,000 tons of D_2O. This is heavy water, in which deuterium replaces the hydrogen, and which is particularly sensitive to certain kinds of neutrinos. Another experiment in the Grand Sasso Tunnel has a very large volume of liquid scintillator, a substance like kerosene. This scintillator material makes flashes of light when hit by neutrinos or other weird particles. Each of these detectors would be especially sensitive if another supernova went off, particularly one in our galaxy. Since the one in the relatively distant Large Magellanic Cloud produced a total of 19 neutrinos in the two detectors that clearly saw it,

The Monopole, Astrophysics, and Cosmic Ray Observatory, one of several detectors under construction at Gran Sasso. MACRO will detect heavy magnetic monopoles and neutral high-energy particles and radiation, such as neutrinos and gamma rays. The apparatus contains thick layers of scintillation counters, each enclosing a sandwich of 10 layers of streamer tubes. The layers of scintillation counters alternate with layers of concrete.

a supernova exploding in our galaxy would produce thousands of neutrinos that could be detected. Detecting neutrinos from such a bright source as a supernova would teach us much about the properties of neutrinos. Since the neutrinos would have certain characteristic time scales and energies, we might be able to learn more about their masses and their properties, including whether or not they can change from one kind into another.

How do we know whether these underground observatories will detect a supernova? We have reasonable estimates that a supernova will occur in our Galaxy from once every 10 years to once every 50 years. It would be very nice if, whenever the next one explodes in our galaxy, we have some underground detectors ready to observe its neutrinos. We will "see" the neutrinos even if the dust in the galaxy blocks all the light of the supernova from reaching us.

Another kind of underground detector that requires special shielding to avoid all other kinds of background noise is a detector to directly observe the dark matter. We mentioned in Chapter 6 that the bulk of the matter of the universe is probably dark (nonshining) stuff not made out of baryons. We don't know what it is. We hope to find some new kind of long-lived, stable particle in future accelerator experiments, and this new particle may be the dark matter. However, another approach to searching for the dark matter is building detectors that can pick up this very low-energy, weakly interacting stuff. It is very difficult, however, to distinguish these signals from the thermal background noise. Several new kinds of low-temperature technologies have been developed to do such searches. Work is still continuing on each approach to make sure that if a signal is seen, it is not something ordinary, but indeed the dark matter. It is hoped that by the 1990s, the backgrounds will have been measured and the technology will be fixed upon, so that a deployment of such dark-matter detectors will be in full swing.

Most of these detectors would operate at very low temperatures, minimizing the noise within the detector. Furthermore, they use relatively large amounts of material, so that even though the probability of interaction is small, at least some interactions will occur. Most of the techniques are aimed at the so-called cold dark matter, the dark matter that may be forming the halo of our galaxy, and that would have been slowly moving at the time galaxies first began to form.

Not everyone believes that cold dark matter is the stuff we're searching for. Another possibility for dark matter is low-mass neutrinos. These low-mass neutrinos would be hard to detect in a dark-matter detector, but measuring a mass for the neutrino of the appropriate size would verify that it is indeed the dark matter. Perhaps the best way to measure the mass of neutrinos, such as the muon or tau neutrino, would be to detect neutrinos from a supernova in our galaxy. Once it was established that the electron, muon, and tau neutrinos had all been detected within a certain time scale, we could use that scale to put limits on the relative masses of the different neutrino types. Thus, there is a close interrelationship between these underground experiments. They all might tell us about the dominant matter of the universe. Furthermore, each of these well-shielded detectors can also tell us about sources in space that might produce extraordinarily high energy neutrinos coming from a set direction. In this way, they would be complementary to the very high energy gamma-ray detectors, since neutrinos, like gamma rays, will not have their directions altered by the galactic magnetic field. There very probably are objects in space that produce high-energy neutrinos, just as there are objects in space that produce high-energy gamma rays. Unlike supernovas, which emit neutrinos for only a few seconds, these as-yet-undiscovered objects might operate in some sort of long-term pulsating way, where the emission comes in discrete bursts that continue sporadically for many years. There are hints of this already from the high-energy gamma-ray sources.

SUMMARY

We have seen in this chapter that there will be a variety of new tools in the next decade that, we hope, will find answers to many of our questions about the fundamental forces and particles of the universe and how they are unified. We hope to understand the origin of the universe, how it is put together, and how galaxies and clusters form. In the course of doing this, we may learn about other exciting objects in space, and about how particles are accelerated in space, which might, in turn, give us ideas about how to build better future accelerators. We also see that these tools for the future are often cross disciplinary. The detector developments from high-energy accelerators are now used in underground detectors or in detectors such as the Fly's Eye, that look for high-energy particles coming from space, thus doing astronomy with particle-physics techniques.

8 EPILOGUE

The types of questions we are now asking and the tools we will use to attempt to answer these questions are truly mind boggling. Only a few years ago, questions such as "Why does the universe have three dimensions (plus time)?" or "What came before the big bang?" were considered philosophical or theological, and not to be pursued within the discipline of physics. Although we may not know the answers, at least the questions have now moved within the grasp of physics. That is, it is now conceivable that the laws of physics, when fully known, will provide satisfying answers to these questions. Before closing this book, we think it will be useful to reflect on the *nature* of the questions we are now asking, rather than on the questions themselves. What are the issues which we want an ultimate "theory of everything" to address?

Many scientific papers have been written about or have been described as being about how a universe can spontaneously be created from "nothing," by arguing that if "nothing" is defined in some mathematical sense to imply an absence of space-time as we know it, then from quantum mechanics (the uncertainty principle) space-time and hence our universe might spontaneously be generated. In particular, these speculations assert that space-time is intrinsically related to the forces and particles, as it is in all unified theories that include gravity. Once some space-time is generated [whether ten-dimensional or (3 + 1)-dimensional is irrelevant here], the whole universe, big bang and all, naturally follows as a consequence of the laws of nature. Quantum mechanics, as the framework in which the universe operates, is presumed rather than derived in current discussions. We know it "works" in the present day, but what is its intrinsic origin? Here it is useful to recall Richard Feynman's comment: "No one understands quantum mechanics."

The question of dimensions is a point of much current research. If space-time starts out in ten dimensions, some aspect of the theory is

A computer simulation of two 20-TeV protons colliding head on. By means of theoretical analysis, physicists have used the known data at 2 TeV to project collision results as 40 TeV—that is, what quarks, gluons, etc., will fly off and convert to pions, protons, kaons, leptons, and other particles.

expected to "compactify" six of them, that is, curl them up to unobservable compactness to obtain a world of three space dimensions and one time dimension. Equivalently, all dimensions were originally compact, but only "3 + 1" were subject to expansion to give the presently observed universe. At first glance, all of this sounds like medieval mystics discussing the music of the spheres, angels on the head of a pin, or some similar early approach to cosmology. Is it just a mathematical game we are playing, is it just semantics, or is it reality that we are seeking?

The approach we've emphasized in this book is that the accepted laws of physics *work!* We carry out experiments that are replicable, and the laws we use can explain the results. But we're probably not going to duplicate the universe in some future experiment (although even this has been discussed in some papers). What we can do is push our experiments so that they duplicate various aspects of the early universe. Similarly, we can check all the predictions of our theories, both in the astronomical domain and in the particle domain, to see if they all fit together in a mathematically consistent manner.

In fact, *mathematical consistency* is really our ultimate goal. Theoretical physicists choose theories for their intrinsic beauty, and this implies consistency, not only consistency with experimental results, but also intrinsic mathematical consistency. Theorists avoid theories with infinities (singularities) or other anomalies that are difficult to deal with in a precise way. It is from such mathematical-consistency arguments that the currently most popular TOE, superstrings, has surged to the top. This superstring theory avoids many of the inconsistencies that other, less complete theories have. Many believe that when the ultimate TOE is found, it will be the only theory that is truly mathematically consistent, with no contradictions, no arbitrary initial conditions, and no anomalies or singularities that yield physical infinities. If such a "Holy Grail" is indeed running the universe, then the true ultimate first cause would be just *mathematical consistency*.

Although mathematical consistency may indeed be the underlying "force" of the universe, to really prove that we have the ultimate theory requires experimental and observational verification. Furthermore, theoretical stumbling blocks or choices among multiple possibilities have in the past been surmounted, with laboratory discoveries pointing the way. Unfortunately, current theories only do their exciting things at extraordinarily high energies, energies far beyond the realm of current experiments. As we've mentioned repeatedly in this book, such high energies did occur in the early universe, as the now-symbiotic relationship between cosmology and fundamental particle physics has demonstrated. But will such cosmological interactions be sufficient to unequivocally "prove" the theory or to provide the insight needed to surmount a barrier? We will return to this point later when we discuss "When will we know it all?"

So far we've discussed our ultimate goal as a single theory of everything, which will be unique—the only theory to be mathematically consistent. Some respectable scientists have taken a somewhat different approach. They have argued that instead of searching for a unique, mathematically consistent theory, the properties of our universe are governed instead by "the anthropic principle." This principle asserts that the laws of physics are what they are because they are the only laws that can produce intelligent beings (us?) who can contemplate the universe. For this principle, the only universes that count are those that ultimately evolve beings who can contemplate them.

Followers of the anthropic principle argue that a huge number of universes might exist with very different physical laws, but the laws in our universe are the only ones that enable intelligent beings to exist. Such a principle is one way to avoid worrying about finding a unique, mathematically consistent theory. Instead, we might assume that there exists a plethora of mathematically consistent theories, but most don't yield people and are no fun.

An alternative to the anthropic principle is what physicists call "naturalness." Naturalness argues that our laws of physics are inevitable. For example, the anthropic principle might argue that there were infinite types of possible initial conditions for the universe, some hot, some cold, and so forth, but only one set of conditions out of an infinite set of possible conditions was able to yield our universe. Advocates for naturalness, however, would argue that some physical mechanism would force the universe to behave the way it has been observed to behave, regardless of the initial conditions.

An example of a successful naturalness mechanism is cosmic inflation. With inflation, even a very irregular, lumpy, anisotropic universe can be made very smooth, homogeneous, and isotropic. Thus a wide variety of initial conditions are made into our observed universe. We don't have to require exceptionally "smooth" initial conditions.

Although there have been entire books written on the anthropic principle (e.g., J. Barrow, and F. Tipler's *The Anthropic Principle,* Oxford University Press, 1986), we really don't see it as very useful. In our opinion it is a way to avoid answering questions ("things are as they are because they were as they were"), rather than a principle that has led to new understanding. On the other hand, the search for "naturalness" has led to new understanding, such as the concept of inflation and the search for a supergrand unified theory of everything that has no free parameters. Although we haven't yet found the latter, the search itself can stimulate rather than stifle physical understanding.

Let us now turn our attention from the esoteric to the pragmatic. In particular, we have mentioned the philosophical aspects of the lofty questions of creation, but physics really progresses by a close interplay with experiment and observation.

When theory gets too far ahead of experiment, the field gets more like philosophy or theology, where debates are difficult to definitively resolve within a human lifetime. When experimental progress gets far ahead of theoretical understanding, we end up with tables and tables of apparently unrelated facts. Disciplines such as classical botany, where facts and details far outreach quantitative understanding, have moved at glacial speeds for decades.

Experimental progress is directly related to technological progress. Because of the need for a close interplay between experiment and theory, there is a symbiotic relationship between technology and science. Sometimes a technological breakthrough enables the experimental exploration of some domain that was previously unexplored. In astronomy, examples of such a domain are the new wavelength regions, such as X rays or radio waves. In particle physics, examples are not only higher energies, but also detector technology that enables more precise measurement of quantities such as position, time, momentum, energy, or decay modes. When new domains are explored, or old domains with new precision, unanticipated discoveries frequently occur. These new discoveries cry out for explanations.

The new explanations can, at times, lead to more general theories that predict new phenomena in other domains that have also not been studied. In this case, theorists then stimulate experimenters to try to develop the technology to reach those domains. And so technology begets science, which, in turn, begets technology . . . .

It should also be noted that this new technology, which science stimulates to explore new domains, frequently has other spin-offs. The excitement of probing the frontiers of scientific understanding attracts some of the world's most technologically gifted people. The new gadgets these people create to understand the universe also may have other applications far beyond their original design. Examples abound, and we will just mention a few. The CAT scan technique in medicine is really an offshoot of nuclear and particle-physics technology. The most efficient solar-energy collectors were first designed as gadgets to optimize detection of showers of elementary particles. The invention of new computer architectures based on massive parallel microprocessors became a necessity for handling particle physics data rates. These will surely have applications in other fields requiring very high speed, special-application computing. Accelerators invented to study quarks are routinely used to provide intense beams of X rays for biologists, chemists, and researchers in condensed matter. Also, a large medical and industrial accelerator industry now exists.

Most of the development of scientific observatories in space was done during the 1970s. During the 1980s, the U.S. space program shifted its effort from doing science in space to developing the space shuttle. For space scientists, the manned program has been frustrating. The few scientific experiments that have been conducted with the shuttle could have been done cheaper and better with unmanned launches. NASA's emphasis on establishing a human presence in space has resulted in a paucity of new space-based observations during the 1980s. Satellites flown by the Europeans, the Japanese, and the Soviets have provided the only new data. With the Challenger disaster has come a reanalysis of NASA's procedures that is yielding a space-science project policy that will once again enable unmanned science experiments to be part of the kind of serious development of astronomical studies that was originally foreseen back in the 1960s and 1970s.

The application of spin-offs of current particle physics and astrophysics now ranges over much of our society's economy. Although the spin-offs, or applications of the results of research, can never be predicted, history has repeatedly shown that basic research eventually leads not only to aesthetic but also to practical effects.

When the British prime minister asked Michael Faraday of what use was his electromagnetic theory, Faraday replied that he did not know, but he was sure the government would one day figure out how to tax it. Since much of our modern lifestyle depends on electromagnetism, it is fair to say, Faraday was right.

In closing this book, let us return to the question, "When will we know it all?" Is the end in sight?

Some fields of science undergo rapid growth, such as now seems to be occurring in particle physics and cosmology. Then they may become relatively stable for centuries, once a certain set of questions is answered. An example is celestial mechanics, following Newton. In other fields, the answer to one question opens up a new class of deeper questions. Particle physics/cosmology has been characterized by this latter kind of ferment until now, but will it always be so? The questions being asked appear more fundamental than ever before. Are they the ultimate questions? Can physicists really find a single theory of everything that is unique and has no free parameters, so that it predicts everything?

The worry here is that our most ambitious theories operate way beyond the energy scales that we can test in accelerators. Thus, at face value this would move the field from the close interplay of experiment and theory to the less resolvable world of philosophy or mathematics. However, we do have the new input from astrophysical observations that may be giving new glimpses of the creation event. There is also the previously mentioned possibility that a mathematical theory may be found that is uniquely self-consistent. For now, this latter possibility still seems far away, although superstring theory advocates are zealously pursuing that avenue.

Our own feeling is that the field will continue to be rich and exciting for decades to come, as accelerators creep up to higher energy scales, and astronomical observations give us broader glimpses of the creation epoch. However, until we really are able to probe Planck-scale energies of 10^{19} GeV per particle in an experimental way, we may not know for sure if we have all aspects of an ultimate theory of everything.

In the 1950s, Stanley Livingston of Lawrence Berkeley Laboratory made an empirical plot of the energy available in laboratory experiments versus time (see the figure on this page). He found a logarithmic relationship in that the energy available increases by roughly a *factor* of 10 every ten years. This relationship has continued to hold into the 1980s with the new colliders such as the Tevatron, and will continue to hold if we can build the SSC within the next decade. If we extrapolate, the Livingston curve predicts we should have the technology to achieve Planck-scale energies by about the year 2150. Skeptics will now surely be outraged by this misuse of the logarithm. Just wait! Obviously, that technology would involve something radically different from present

The Livingston curve has been an accurate empirical estimator of the energies that accelerators have been able to explore for the entire twentieth century. If we presume that technology will continue to improve in the next two centuries at this same rate, then we might achieve Planck-scale energies in laboratory experiments by the year 2150. Such experiments might even produce new inflating universes. (Maybe we're the product of some previous universe's Planck-scale experiments?)

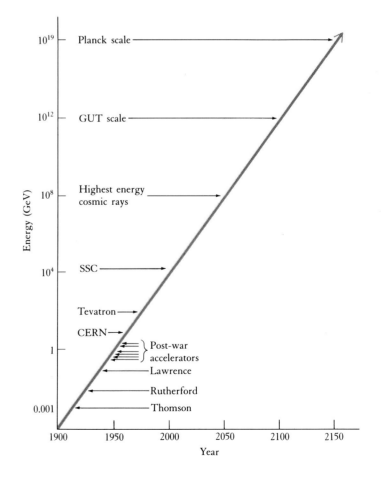

technology, since scaling the Tevatron up to Planck energies would require an accelerator of galactic dimensions. However, the Livingston curve has worked in the past, with new technology, such as colliders, coming along just in time to enable us to continue to reach the higher energies predicted by the relationship.

Remember that, although 10^{19} GeV per particle is a huge energy by all current accelerator standards, when viewed in macroscopic terms, it is only the kinetic energy of a Lear jet. To make the accelerator of the year 2150, we need to repeatedly put the energy of a Lear jet into a single proton. This is an awesome task for our progeny, because in order to learn something, we must involve a reasonable number of particles. Today's weaker beams have approximately 10^6 particles—a lot of Lear jets!

It is more likely that by 2150 we may find that the new questions require energies beyond the Planck scale. Or, even more likely, we may have found that a mathematically consistent theory has explained all results from all new accelerators, from the "big accelerator in the sky"— the early universe—and has accounted for all the details of the standard model. In that case, there is no need to build the 2150 machine. Finally, we must admit the possibility that neurobiology will usurp all the research funds and theorists will don a helmet in lieu of seeking the TOE.

Rather than speculate further on life two centuries from now, let us instead refocus on the current need to continue the progress on both fronts: inner space and outer space. This requires the Hubble Telescope and the other powerful tools planned for the 1990s, including the SSC. Unless we can resolve the current questions about unification, supersymmetry, and the existence of some vacuum energy, we have no hope of pushing on toward a true theory of everything. With the SSC, we should be able to resolve our current primary questions, and also show that our society is able to keep up with the Livingston relation. Since this is a projection from experience, it clearly must be correct and so it is a chart of splendid optimism. It tells us we will have solved the energy crisis, as well as the environmental, fiscal, and political problems that so obsess us at the end of the twentieth century. Above all, it predicts we shall survive as a civilized society! And what about the theory of everything?

When we achieve this TOE, elegant, simple, fitting comfortably on campus T-shirts, even then the essential element that must be preserved is the possibility that it is subject to change, that it can be improved. For it is this salient attitude that defines science.

SOURCES OF ILLUSTRATIONS

Drawings by George Kelvin and Vantage Art, Inc.

Facing page 1
Anglo-Australian Observatory

Page 2
Fermi National Accelerator Laboratory

Page 4
Fermi National Accelerator Laboratory

Page 5
Fermi National Accelerator Laboratory

Page 6
Joe Stancmpiano, NGS Staff and Karl Luttrell, University of Michigan © 1988 National Geographic Society.

Page 7
Anglo-Australian Observatory

Page 9
Fermi National Accelerator Laboratory

Page 12
IBM Almaden Research Center/Critical Review Letter vol. 60, #23, p. 2398, 1988.

Page 15
Dennis di Cicco

Page 16
Anglo-Australian Observatory

Page 18
Larry Smarr, David Hobill, David Bernstein, Ray Idaszak, and Donna Cox/National Center for Supercomputing Applications.

Page 20
Giuseppe Bezzuoli, Galileo dimostra la leggedella cuduta dei gravi, Firenze, Museo Zoologico: Scala/Art Resource

Page 22
Adler Planetarium, Chicago

Page 23 (top)
The Science Museum, London

Page 24
Berenice Abbott/Commerce Graphics Ltd., Inc.

Page 27
The Exploratorium

Page 28
Deutsches Museum

Page 29
Cavendish Laboratory

Page 32
Chip Clark

Page 45
Larry Smarr, David Hobill, David Bernstein, Ray Idaszak, and Donna Cox/National Center for Supercomputing Applications.

Page 48
CERN

Page 51
The Science Museum, London

Page 52
The Science Museum, London

Page 57
Fermi National Accelerator Laboratory

Page 62
U.S. Dept. of Energy

Page 65
Lawrence Berkeley Laboratory

Page 66
The Archives, California Institute of Technology

Page 67
from "Physical Review" 43, 491 (1933).

Page 73 (top)
George D. Rochester

Page 73 (bottom)
from *The Study of Elementary Particle Physics by the Photographic Method* C. F. Powell, P. H. Fowler, and D. H. Perkins. New York: Pergamon Press, 1959, p. 245.

Page 75 (left)
CERN

Page 75 (right)
Fermi National Accelerator Laboratory

Page 77
Fermi National Accelerator Laboratory

Page 78
CERN

Page 81
Anglo-Australian Observatory

Page 84
Lawrence Berkeley Laboratory

Page 90
CERN

Page 93
Lawrence Berkeley Laboratory

Page 94
Lawrence Berkeley Laboratory

Page 96 (top)
Brookhaven National Laboratory

Page 101
Brookhaven National Laboratory

Page 103
Stanford Linear Accelerator Center and the
U.S. Dept. of Energy

Page 104
Fermi National Accelerator Laboratory

Page 105
Deutsches Elektronen-Synchrotron

Page 106
Lawrence Berkeley Laboratory

Page 112
Deutsches Elektronen-Synchrotron

Page 116
Fermi National Accelerator Laboratory

Page 117
Fermi National Accelerator Laboratory

Page 118 (left)
Fermi National Accelerator Laboratory

Page 118 (right)
Fermi National Accelerator Laboratory

Page 128
National Radio Astronomy Observatory

Page 130 (left)
Yerkes Observatory

Page 130 (right)
National Optical Astronomy Observatories

Page 132
Tektronix, Inc.

Page 133 (bottom)
National Optical Astronomy Observatories

Page 138
AT&T Archives

Page 139
F. N. Owen and J. J. Puschell/National
Radio Astronomy Observatory

Page 142
AT&T Archives

Page 143
The National Astronomy and Ionosphere
Center, operated by Cornell University
under contract with the National Science
Foundation.

Page 146 (top)
National Optical Astronomy Observatories

Page 146 (middle)
J. O. Burns, E. J. Schreier, and E. D.
Feigelson/National Radio Astronomy Ob-
servatory

Page 146 (bottom)
C. Jones, C. Stern, and W. Forman/Smith-
sonian Astrophysical Observatory

Page 149
NASA

Page 158
Adrian Melott, University of Kansas, and
Sergei Shandarin, Institute for Physical
Problems, Moscow, USSR

Page 163
NASA

Page 168
Anglo-Australian Observatory

Page 170
P. J. E. Peebles

Page 178
Adrian Melott, University of Kansas, and
Sergei Shandarin, Institute for Physical
Problems, Moscow, USSR

Page 179
M. J. Geller, J. P. Huchra, V. de Lapparent,
and R. McMahan, Harvard-Smithsonian
Center for Astrophysics.

Page 182
Kamiokande-II Collaboration

Page 183
Chiolini

Page 188
California Association for Research in
Astronomy

Page 194
CERN

Page 199
Brookhaven National Laboratory

Page 201
SSC Central Design Group, Lawrence
Berkeley Laboratory

Page 203
Fermi National Accelerator Laboratory

Page 205
Fermi National Accelerator Laboratory

Page 208
MMT Observatory

Page 210
European Southern Observatory

Page 214
Cosmic Ray Physics Dept., University of
Utah

Page 217
Lockheed Missiles & Space Comp., Inc.

Page 219
NASA

Page 221
TRW Inc.

Page 223
L. Di Giuseppe

Page 226
Fermi National Accelerator Laboratory

INDEX

Other books in the Scientific American Library Series

POWERS OF TEN
by Philip and Phylis Morrison and
the Office of Charles and Ray Eames

HUMAN DIVERSITY
by Richard Lewontin

THE DISCOVERY OF SUBATOMIC PARTICLES
by Steven Weinberg

THE SCIENCE OF MUSICAL SOUND
by John R. Pierce

FOSSILS AND THE HISTORY OF LIFE
by George Gaylord Simpson

THE SOLAR SYSTEM
by Roman Smoluchowski

ON SIZE AND LIFE
by Thomas A. McMahon and John Tyler Bonner

PERCEPTION
by Irvin Rock

CONSTRUCTING THE UNIVERSE
by David Layzer

THE SECOND LAW
by P. W. Atkins

THE LIVING CELL, VOLUMES I AND II
by Christian de Duve

MATHEMATICS AND OPTIMAL FORM
by Stefan Hildebrandt and Anthony Tromba

FIRE
by John W. Lyons

SUN AND EARTH
by Herbert Friedman

EINSTEIN'S LEGACY
by Julian Schwinger

ISLANDS
by H. William Menard

DRUGS AND THE BRAIN
by Solomon H. Snyder

THE TIMING OF BIOLOGICAL CLOCKS
by Arthur T. Winfree

EXTINCTION
by Steven M. Stanley

MOLECULES
by P. W. Atkins

EYE, BRAIN, AND VISION
by David H. Hubel

THE SCIENCE OF STRUCTURES AND MATERIALS
by J. E. Gordon

SAND
by Raymond Siever

THE HONEY BEE
by James L. Gould and Carol Grant Gould

ANIMAL NAVIGATION
by Talbot H. Waterman

SLEEP
by J. Allan Hobson